Cancer Enzymology

MIAMI WINTER SYMPOSIA

1. W. J. Whelan and J. Schultz, editors: HOMOLOGIES IN ENZYMES AND METABOLIC PATHWAYS and METABOLIC ALTERATIONS IN CANCER, 1970

2. D. W. Ribbons, J. F. Woessner, Jr., and J. Schultz, editors: NUCLEIC ACID-PROTEIN INTERACTIONS and NUCLEIC ACID SYNTHESIS IN VIRAL INFECTION, 1971

3. J. F. Woessner, Jr. and F. Huijing, editors: THE MOLECULAR BASIS OF BIOLOGICAL TRANSPORT, 1972

4. J. Schultz and B. F. Cameron, editors: THE MOLECULAR BASIS OF ELECTRON TRANSPORT, 1972

5. F. Huijing and E. Y. C. Lee, editors: PROTEIN PHOSPHORYLATION IN CONTROL MECHANISMS, 1973

6. J. Schultz and H. G. Gratzner, editors: THE ROLE OF CYCLIC NUCLEOTIDES IN CARCINOGENESIS, 1973

7. E. Y. C. Lee and E. E. Smith, editors: BIOLOGY AND CHEMISTRY OF EUCARYOTIC CELL SURFACES, 1974

8. J. Schultz and R. Block, editors: MEMBRANE TRANSFORMATION IN NEOPLASIA, 1974

9. E. E. Smith and D. W. Ribbons, editors: MOLECULAR APPROACHES TO IMMUNOLOGY, 1975

10. J. Schultz and R. C. Leif, editors: CRITICAL FACTORS IN CANCER IMMUNOLOGY, 1975

11. D. W. Ribbons and K. Brew, editors: PROTEOLYSIS AND PHYSIOLOGICAL REGULATION, 1976

12. J. Schultz and F. Ahmad, editors: CANCER ENZYMOLOGY, 1976

MIAMI WINTER SYMPOSIA - VOLUME 12

Cancer Enzymology

edited by

J. Schultz

F. Ahmad

PAPANICOLAOU CANCER RESEARCH INSTITUTE

MIAMI, FLORIDA

Proceedings of the Miami Winter Symposia, January 1976
Sponsored by The Papanicolaou Cancer Research Institute
Miami, Florida

Academic Press, Inc.   New York   San Francisco   London 1976
A Subsidiary of Harcourt Brace Jovanovich, Publishers

COPYRIGHT © 1976, BY ACADEMIC PRESS, INC.
ALL RIGHTS RESERVED.
NO PART OF THIS PUBLICATION MAY BE REPRODUCED OR
TRANSMITTED IN ANY FORM OR BY ANY MEANS, ELECTRONIC
OR MECHANICAL, INCLUDING PHOTOCOPY, RECORDING, OR ANY
INFORMATION STORAGE AND RETRIEVAL SYSTEM, WITHOUT
PERMISSION IN WRITING FROM THE PUBLISHER.

ACADEMIC PRESS, INC.
111 Fifth Avenue, New York, New York 10003

United Kingdom Edition published by
ACADEMIC PRESS, INC. (LONDON) LTD.
24/28 Oval Road, London NW1

LIBRARY OF CONGRESS CATALOG CARD NUMBER: 76–20852

ISBN 0–12–632745–9

PRINTED IN THE UNITED STATES OF AMERICA

CONTENTS

SPEAKERS, CHAIRMEN AND DISCUSSANTS   ix
PREFACE   xi

Protease-Related Effects in Normal and Transformed Cells 1
   J.M. Buchanan, L. Bo Chen, and B.R. Zetter
      DISCUSSION: Magnusson

The Secretion of Fibrinolysin by Cultured Mammalian Tumor Cells . . . 25
   M.C. Wu, D.R. Schultz, and A.A. Yunis
      DISCUSSION: Brown, Buchanan, and Magnusson

Isozyme Composition, Gene Regulation, and Metabolism
of Experimental Hepatomas. 41
   S. Weinhouse, J.B. Shatton, and H.P. Morris
      DISCUSSION: Chance, Kaplan, Parks, Horowitz,
                        Grossman, and Huijing

Enzymatic Strategy of the Cancer Cell. 63
   G. Weber
      DISCUSSION: Menahan, Grossman, Morris, Koch
                        Parks, Chance, and Buchanan

Oxygen Reduction by Cytochrome Oxidase — A Possible
Source of Carcinogenic Radical Intermediates 89
   B. Chance
      DISCUSSION: Weber, Theorell, Vallee, Estabrook
                        Kaplan, Kareem, and Mildvan

The Activation of Polycyclic Hydrocarbons:
Cytochrome's P-450, Oxygen and Electrons. 103
   R.W. Estabrook, V.W. Patrizi, and R. Prough
      DISCUSSION: Chance, Weinhouse, Mildvan,
                        Vallee, Parks, and Jakoby

CONTENTS

Magnetic Resonance Studies of the Mechanism of
DNA Polymerase I from *E. Coli* 123
    A.S. Mildvan, D.L. Sloan, C.F. Springgate, and L.A. Loeb
        DISCUSSION: *Vallee, Estabrook, Greer, Parks, and Block*

Enzymes Involved in Repair of DNA Damaged by
Chemical Carcinogens and γ-Irradiation 139
    D.M. Kirtikar, J.P. Kuebler, A. Dipple, and D.A. Goldthwait
        DISCUSSION: *Van Lancker, Grossman, and Greer*

Zinc Biochemistry in the Normal and Neoplastic Growth Processes. . . . 159
    B.L. Vallee
        DISCUSSION: *Brada, Mildvan, Weinhouse, Schultz,*
                          Bade, Petering, and Chmielewicz

Target Directed Cancer Chemotherapeutical Agents. 201
    N.O. Kaplan
        DISCUSSION: *Leif, Weinhouse, Menahan, and Parks*

Regulation of Fatty Acid Biosynthesis in Mammary Tumors 229
    F. Ahmad, P. Ahmad, and D. Schildknecht
        DISCUSSION: *Lynen, Weinhouse, Cameron, and Rouleau*

Selective Inhibition of the 3' to 5' Exonuclease Activity
Associated with Mammalian DNA Polymerase δ 245
    J.J. Byrnes, K.M. Downey, V. Black, L. Esserman, and A.G. So
        DISCUSSION: *Mildvan and Koch*

Myeloperoxidase-Mediated Cytotoxicity 267
    S.J. Klebanoff, R.A. Clark, and H. Rosen
        DISCUSSION: *Schultz, Mildvan, Leif, and Estabrook*

Cytotoxicity of the Superoxide Free Radical 289
    J.M. McCord and M.L. Salin
        DISCUSSION: *Harrison, Estabrook, and Horowitz*

The Functional Mechanism of Myeloperoxidase 305
    J. Harrison
        DISCUSSION: *Weinhouse and Klebanoff*

Myeloperoxidase-Enzyme Therapy on Rat Mammary Tumors. 319
    J. Schultz, A. Baker, and B. Tucker
        DISCUSSION: *Mildvan and Leif*

Free Communications

Defective Regulation of Cholesterol Biosynthesis in
Tumor-Virus Transformed and Hypercholesterolemic
Human Skin Fibroblasts: A Comparative Study 335
   J.M. Bailey, T. Allan, E.J. Butler and J.D. Wu

Viral Stimulation of Choline Phosphotransferase in
Spleen Microsomes During Production of Malignancy 336
   W.E. Cornatzer, D.R. Hoffman and D. Skurdal

L-Asparaginase with Antilymphoma Activity From
Vibrio succinogenes 337
   J.A. Distasio and R.A. Niederman

Malate-Aspartate Shuttle Activity in Several Ascites Tumor Lines 338
   W.V.V. Greenhouse and A.L. Lehninger

Inhibition by Serum of Intracellular Degradation of Human
Chorionic Gonadotropin (hCG) 339
   R.O. Hussa and R.A. Pattillo

The Interaction of Antitumor Drugs with Folate Requiring Enzymes ... 340
   D.W. Jayme, P.M. Kumar, N.A. Rao, J.A. North, and J.H. Mangum

Studies on the Collagenolytic Activity of Methylcholanthrene-
Induced Fibrosarcomas in Mice 341
   K.R. Labrosse and I.E. Liener

Activities of Enzymes of Glycolysis and Adenine Nucleotide
Metabolism During Murine Leukemogenesis 342
   L.A. Menahan and R.G. Kemp

The Association of a Protease (Plasminogen Activator) with a
Specific Membrane Fraction Isolated From Transformed Cells 343
   J.P. Quigley

Structural Features of S. Typhimurium Lipopolysaccharide (LPS)
Required for Activation of Monocyte Tissue Factor 344
   F.R. Rickles and P.D. Rick

Role of Plasminogen Activator in Generation of MIF-Like
Activity by SV3T3 Cells 345
   R.O. Roblin, M.E. Hammond, P.H. Black, and H.F. Dvorak

Proteolysis and Cyclic AMP Levels in Cell Culture 346
   W.L. Ryan, M.L. Heidrick, and G.L. Curtis

Specificity of DNA-Dependent RNA Polymerase Activities
in Rabbit Bone Marrow Erythroid Cell Nuclei 347
   M.K. Song and J.A. Hunt

SPEAKERS, CHAIRMEN, AND DISCUSSANTS

F. **Ahmad**, Papanicolaou Cancer Research Institute, Miami, Florida.

M. *Bade*, Boston College, Chestnut Hill, Massachusetts.

R.E. *Block*, Papanicolaou Cancer Research Institute, Miami, Florida.

Z. *Brada*, Papanicolaou Cancer Research Institute, Miami, Florida.

A.M. *Brown*, National Cancer Institute, Rockville, Maryland.

J.M. **Buchanan** (Session Chairman), Massachusetts Institute of Technology, Cambridge, Massachusetts.

J.J. **Byrnes**, University of Miami School of Medicine, Miami, Florida.

B.F. *Cameron*, Papanicolaou Cancer Research Institute, Miami, Florida.

B. **Chance** (Session Chairman), University of Pennsylvania, Philadelphia, Pennsylvania.

Z.F. *Chmielewicz*, State University of New York, Buffalo, New York.

R.W. **Estabrook** (Session Chairman), University of Texas, Dallas, Texas.

D.A. **Goldthwait**, Case Western Reserve University, Cleveland, Ohio.

S.B. *Greer*, University of Miami, Miami, Florida.

S. *Grossman*, Union Carbide Corporation, Tarrytown, New Jersey.

J.E. **Harrison**, Papanicolaou Cancer Research Institute, Miami, Florida.

M. *Horowitz*, New York Medical College, Mt. Vernon, New York.

F. *Huijing*, University of Miami, Miami, Florida.

W. *Jakoby*, National Institutes of Health, Bethesda, Maryland.

SPEAKERS, CHAIRMEN, AND DISCUSSANTS

N.O. Kaplan (Session Chairman), University of California, San Diego, California.

H. Kareem, University of Miami, Miami, Florida.

S. Klebanoff, University of Washington, Seattle, Washington.

G. Koch, Roche Institute of Molecular Biology, Nutley, New Jersey.

R. Leif, Papanicolaou Cancer Research Institute, Miami, Florida.

F. Lynen, Max-Planck-Institute für Biochemie, Bei Munchen.

S. Magnusson, University of Aarhus, Aarhus.

J. McCord, Duke University, Durham, North Carolina.

L. Menahan, Medical College of Wisconsin, Milwaukee, Wisconsin.

A. Mildvan (Session Chairman), Institute for Cancer Research, Philadelphia, Pennsylvania.

H.P. Morris, Howard University, Washington, D.C.

R.E. Parks, Brown University, Providence, Rhode Island.

H. Petering, University of Cincinnati, Cincinnati, Ohio.

M. Rouleau, National Institutes of Health, Bethesda, Maryland.

J. Schultz, Papanicolaou Cancer Research Institute, Miami, Florida.

A.H. T. Theorell, Karolinska Institutet, Stockholm, Sweden.

B.L. Vallee (Session Chairman), Harvard University, Boston, Massachusetts.

J. Van Lancker, University of California, Los Angeles, California,

G. Weber (Session Chairman), Indiana University, Indianapolis, Indiana.

S. Weinhouse (Session Chairman), Temple University, Philadelphia, Pennsylvania.

A.A. Yunis, University of Miami School of Medicine, Miami, Florida.

PREFACE

In recent years a great deal of progress has been made on the covalent modifications of proteins and how they affect physiological function. One of these modifications that profoundly affects many cellular functions involves proteolysis. Therefore, the main theme of the Eighth Miami Winter Symposia arranged by the Department of Biological Chemistry, University of Miami, concerns "Proteolysis and Physiological Regulation." As in the past, the program arranged by the Papanicolaou Cancer Research Institute is complimentary to this basic theme. This year we chose "Cancer Enzymology."

The selection of Dr. Hugo Theorell as the Lynen Lecturer made it possible for us also to select old friends and students of Dr. Theorell: namely, John Buchanan and Britton Chance. It also brought back to the symposia previous participants such as Sidney Weinhouse, Ronald Estabrook, George Weber, and Seymour Klebanoff. The plan of the program started with the strategy of the cell and progressed to bioenergetics, regulation and cytotoxicity, thus providing a role for enzymes in the mechanism of the living as well as protection against adverse environment.

It has always been a pleasure to meet with those who have attended the symposia in the past and continue to do so, each time exercising and expressing their pleasure at the format and general atmosphere of this Gordon Conference-like meeting taking place in mid-winter. The restful atmosphere in Miami at this time of year is conducive to more relaxed discussion and interaction among the participants. Such interaction is not readily possible at large national meetings where both the subject matter and the number of participants limits the opportunity to meet informally and enjoy the criticism of objective observers. It is, therefore, our sincere hope that future Miami Winter Symposia will be attended by many former participants, as well as many new scientists, to share in a mutually stimulating week of academic exchange.

We would like to acknowledge the support of Eli Lilly and Company, Abbott Laboratories, Hoffman-LaRoche, Inc., MC/B Manufacturing Chemists, and the Upjohn Company. Finally, it is of particular importance that the recording of the discussions and the preparation of this volume was made possible through the combined efforts of our expert typists, Kathi Bishop, Bonnie Tracy, and Anne Johnson, who worked under the direction of Mrs. Ginny Salisbury.

J. Schultz
F. Ahmad

Cancer Enzymology

PROTEASE-RELATED EFFECTS IN NORMAL AND TRANSFORMED CELLS

JOHN M. BUCHANAN, LAN BO CHEN AND BRUCE R. ZETTER
Department of Biology
Massachusetts Institute of Technology
Cambridge, Massachusetts 02139

INTRODUCTION

The basic problem of cell biology concerned with the mechanism of stimulation of resting or dormant cells to proliferation has been approached in several ways. There is now a relatively long list of compounds that have been shown to be mitogenic for certain cell systems. This list includes serum (1-3), the protein hormones (4), insulin and insulin-like materials, such as multiplication stimulating activity (5) and somatomedin (6), fibroblast growth factor (7) and epidermal growth factor (8). Other biological materials including fetuin (9), seromucoid (10), lipopolysaccharide (11), glucocorticoids (12) and lectins (11) have also been reported to be mitogenic.

A third group of mitogenic substances, the proteases, have been of particular interest to us since the observation by Sefton and Rubin (13) that trypsin stimulates resting chick embryo cells to division and the report by Burger (14) in the same year that trypsin, ficin, and pronase can stimulate the growth of density-inhibited 3T3 cells. At the time of initiation of studies in our laboratory the role of proteases as mitogens had not been generally explored in many cell systems.

The use of proteases as mitogens seemed particularly attractive initially because they had been reported to react with proteins on the cell surface causing specific alterations that have been proposed to have special significance in the basic mechanism of the processes leading to cell division (15,16). Specifically, a cell surface protein of 250,000 daltons that can be enzymatically iodinated is removed by trypsin (17-20) and is absent from the external surface of many transformed cells (17-25). Furthermore, a protease capable of activating plasminogen has been indirectly implicated in maintaining the phenotype of transformed cells (26-30). Thus, proteases have been proposed as active agents in the metabolism of both normal and transformed cells.

At the onset of our experiments we felt that certain restrictions should be applied to the choice of a protease selected for study.

It should

1) be a major constituent of the circulating fluids normally found in blood, and therefore in plasma, as a proenzyme, which can be activated if the need for cell proliferation occurs,

2) be readily available in relatively large quantities in highly purified form,

3) exhibit a high degree of proteolytic specificity,

4) serve as a mitogen for a host cell that will respond to this enzyme when present alone without the requirement of any other survival, attachment or growth factors.

Of the several possible choices among the kallikreins and the proteases of the clotting and complement systems, thrombin was selected as most nearly fulfilling these qualifications. Its proenzyme form, prothrombin, is present in blood at a level of 100 γ per ml. Serum contains active thrombin since the latter is produced in excess of the capacity of antithrombin proteins to neutralize it (31). The relationship of thrombin to other blood proteases is shown in Fig. 1.

Thrombin is a highly specific proteolytic enzyme that is known to split, for example, 4 arginyl-glycine bonds of fibrinogen (32) and a few selected lysyl and arginyl bonds of actin (33). It has been supplied to us in highly pure form from bovine and human sources by Dr. David F. Waugh and Dr. John W. Fenton, respectively.

RESULTS

Thrombin as a mitogen:

When added as a single component to the culture medium

of chick embryo fibroblasts, thrombin is capable of stimulating these cells to division (34). The assay for mitogenicity consists of measuring the response of resting cells for DNA synthesis from labeled thymidine 12 hr after addition of the reagent or after 4 days for total cell count.

By both criteria thrombin was highly mitogenic. When compared to graded levels of serum ranging from 0 to 5 percent, thrombin was equally effective when added between 0 and 10 μg per ml. Prothrombin, itself, was inactive but when incubated in a Ca^{++}-containing medium to which Factor X_a and V had been added in very small quantities, the proenzyme was converted to the active agent, thrombin. In this conversion the chick cells themselves provided the thromboplastin, a colloidal lipoprotein necessary for this reaction (see Fig. 1).

We then explored the relative proportions of thrombin-yielding components of defibrinogenated plasma to other mitogenic activities as calibrated by use of the chick embryo fibroblast as the host cell. The answer to this question was first approached by determination of the inhibition of the mitogenic activity of thromboplastin-treated plasma with phenylmethylsulfonylfluoride, which combines with and inactivates the "serine" proteases (32). Approximately 50% of the mitogenic activity of plasma was lost by this procedure. However, treated-plasma may contain other "serine" proteases with mitogenic activity, but in far less quantities than thrombin. One example is Factor X_a, which has been shown in this laboratory to be mitogenic (unpublished results).

Therefore, in order to assess the true contribution of prothrombin in plasma we separated the proteins in de-fibrinogenated plasma that are absorbed to $BaSO_4$ and tested the mitogenic activities of both the non-absorbed (Fraction I) and absorbed proteins (Fraction II) before and after treatment with thromboplastin. It was found that the total mitogenic activity of Fraction I did not depend on treatment of the preparation with thromboplastin, but that the activity of Fraction II was greatly enhanced by this treatment. When chromatography was performed on either the untreated or treated samples of Fraction II, we could show that in the former case the mitogenic activity was eluted with prothrombin and in the latter instance with thrombin. The mitogenic activity from thrombin precursors of plasma

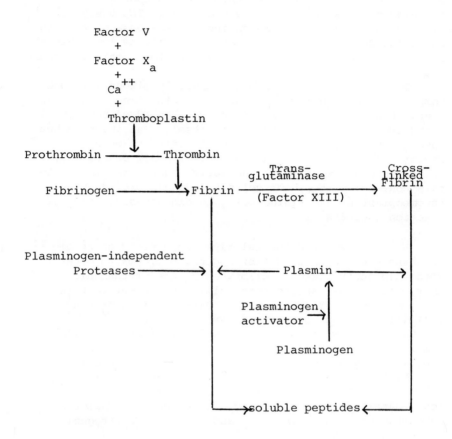

Fig. 1. Enzyme systems involved in the formation and dissolution of fibrin clots.

is approximately 30 percent of the total mitogenic activity. Thus, by either method of analysis, i.e. by specific analysis or by enzymatic inhibition with phenylmethylsulfonylfluoride, the thrombin-yielding material of plasma represents an important share of the mitogenic activity of this fluid. Therefore, from a quantitative point of view thrombin formed from prothrombin might be expected to play a role under certain physiological conditions, for example in wound healing. The concentration of thrombin in the microenvironment of the wound during clot formation probably is considerably greater than that of collected serum and would be sufficient to initiate cell proliferation in the wound area, a necessary second step in restoring the damaged area to normal condition.

Since these experiments were conducted entirely with secondary cultures of chick embryo fibroblasts, we then examined the role of thrombin in other cell systems. When added to several murine embryo fibroblasts (mouse-Swiss, mouse BalB C, Rat, and Golden Hamster) thrombin was also strongly mitogenic. In a very surprising manner it could not under these conditions cause cell proliferation when tested on chick embryo fibroblasts transformed with Rous sarcoma virus. Whether this lack of response to thrombin represents a requirement for other serum factors or whether the transformed cell simply could not respond to the protease as a growth stimulator required further investigation.

There are certain cell lines that are unresponsive to thrombin when the latter is present as the sole mitogen, for example, 3T3, 3T6, CHO, BHK and Nil. In recent experiments we have examined the effects of thrombin on some of these cells when it is added in combination with other mitogenic factors.

In Fig. 2 we show results of an experiment with a mouse BalB C, 3T3 fibroblast cell line infected with Avian sarcoma virus B77. As the accessory factors we selected Cohen's epidermal growth factor (8) and prostaglandin F2α. The latter has been demonstrated to have a stimulatory effect on DNA synthesis and cell proliferation of Swiss 3T3 cells (35). The B77-3T3 line cell was used in these experiments because we have found that it is particularly responsive to epidermal growth factor. The cell count in all cases was normalized to the growth response obtained with 5% calf serum.

At a low concentration of serum (0.7%) B77-3T3 is maintained in culture but exhibits only minimal growth. When provided with thrombin at a level of 0.2 µg/ml there was no growth beyond this basic level. Highly purified epidermal growth factor supplied at a level of 1.5 ng/ml evinces a response 1.4 times that of the 5% serum level. At a level of 0.6 ng/ml the response is only 50%. It is at this latter concentration that effects of other factors can be demonstrated. For example, addition of prostaglandin F2α at

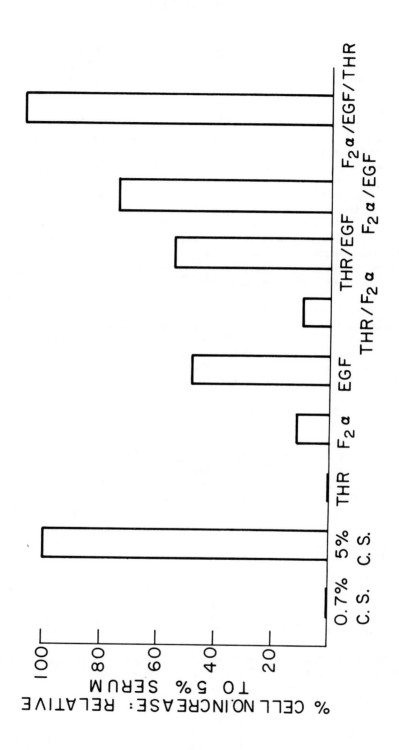

Fig. 2. Potentiation of cell growth of B77-3T3 mouse fibroblasts by thrombin in the presence of other co-mitogens. EGF=epidermal growth factor (0.6 ng/ml) Thr=thrombin (0.2 µg/ml), F2α= prostaglandin F2α (0.2 µg/ml), C.S. = calf serum.

a concentration of 0.2 µg/ml to the B77-3T3 cells results in an increase of only 10% of that of 5% serum. Combination of thrombin with either F2α or EGF does not yield a further stimulation of cell division. However, combinations of F2α and EGF show slightly more than the additive effects of either alone, but in the presence of all three factors, F2α, EGF and thrombin, a response greater than that evoked with 5% serum is obtained. Therefore, it is possible to demonstrate a role for thrombin provided that other essential factors are present. This potentiation for thrombin cannot be demonstrated, however, at the higher saturating concentrations of EGF. Similar synergistic effects of EGF and arginine esterase (or thrombin) on the stimulation of DNA synthesis in an explant of human foreskin has been independently observed by Lembach (36).

The Role of Surface Proteins in Cell Proliferation as Studied with Thrombin and Other Proteases:

As stated in the introductory comments, special attention has centered on an iodinatable cell surface protein of 250,000 daltons (250K) of chick embryo fibroblasts, since it is removed by trypsin during mitogenic stimulation and is lost by some transformed cells. We have attempted to provide further evidence about the role of this protein in mitogenesis by use of thrombin (37). Thrombin is ideally suited for this type of investigation because of its mitogenicity and because of its great specificity as demonstrated by the limited number of peptide bonds it can split. In addition to trypsin and thrombin, several other proteolytic enzymes including bromelin, ficin, chymotrypsin, subtilisin, pronase, and α-protease, were tested both for their capacity to stimulate cell proliferation (or DNA synthesis) and to cleave iodinatable cell surface proteins of chick embryo fibroblasts. Our first efforts centered around an attempt to establish a correlation between mitogenicity and cell surface composition after protease treatment. Two comparisons stood out as particularly revealing, namely, that thrombin could stimulate cells to divide, yet did not cleave the 250K protein, whereas chymotrypsin was not mitogenic but cleaved this cell surface component. These facts by themselves indicated fairly clearly that protease-mediated loss of the 250K protein is not an event sufficient by itself to bring about cell division nor is its removal necessary for cell proliferation. The results of these studies required a reassessment of previous speculations on the role of specific membrane proteins in the regulation of growth in normal cells and probably also in transformed cells.

In spite of the failure of these experiments to implicate the 250K protein, it seemed likely that thrombin was having its effect on the chick embryo fibroblasts by virtue of its proteolytic activity. Close examination of the autoradiograms of gel strips of the chick fibroblast surface proteins iodinated with ^{125}I suggested that an iodinated protein slightly smaller than the 250K protein was lost after thrombin treatment. By modification of the procedure for detection of cell surface components, namely by substitution ^{131}I for ^{125}I, it has been possible to increase greatly the resolution of bands of iodinated proteins formed during electrophoresis on gradient polyacrylamide gel slabs. Under these conditions a protein of approximately 205,000 daltons can easily be detected as a prominent cell surface component (38). An examination of the fate of this protein after treatment of the cells separately with several proteases including thrombin showed that without exception it was lost. This observation at first seemed to provide the clue to the function of proteases during their interaction with cells until it was appreciated that in certain cases, for example, treatment with either α protease, thermolysin or papain, there was no mitogenic stimulation (39). Thus, removal of the 205K protein is not a sufficient condition for the mitogenic response elicited by some proteases. However, we have as yet no further evidence that it may not be a necessary condition.

It should be pointed out that in all of the experiments so far cited serum was excluded from all of the plates that were treated with proteases, so that the participation of an added co-mitogen from this source was ruled out in the case of these particular cells.

What, then, is the effect of alteration of the cell surface proteins of chick embryo fibroblasts by certain proteases if stimulation of cells to division does not correlate with their presence or absence?

One property that does seem to correlate with the composition of the surface proteins is the cell morphology. Chick embryo fibroblasts grown in serum assume a parallel arrangement when present in the confluent state. On the other hand, protease-treated cells seem to lose this characteristic pattern and in some cases, notably with thrombin, the cells migrate towards each other to form large clumps with conspicuous empty spaces between these formations. This pattern of cell distribution was observed for fibroblasts treated with thrombin, α protease or bromelin, but not by four cell cultures, treated either with thermolysin, trypsin, papain or elastase (39).

Upon inspecting our data, (Table 1) we noted that in the first instance, the protease was capable of removing the 205K protein but not the 250K protein and in the second instance of four proteases, both cell proteins were removed (39). At this juncture, a paper by Yamada, Yamada and Pastan (40) appeared in which they reported that addition of purified 250K (or LETS) protein to sheep erythrocytes caused them to agglutinate. They postulated that the function of the LETS protein in fibroblast cultures was involved in cell adhesion. Our results provide direct evidence for this postulation. We propose (39) that protease treatment of chick embryo fibroblasts consistently brings about two responses; removal of a 205,000 dalton protein from the cell surface and an initial morphological change characterized by reduction of both cytological spreading and cell-cell interactions. In some cases, namely in those instances where the 250K protein remained intact, this first event is followed by a later independent event involving increased migration and aggregation of cells.

Proteases and the Transformed Cell:

Aside from their putative importance in the physiological processes of the normal cell, proteases are believed to play an important role in the growth properties of the transformed cell. Attention has been focused on the latter relationship by Reich, Rifkin and their colleagues (15,26-20). Starting with the observation by Fischer (41) that tumors observed in culture are surrounded by partially dissolved plasma clots, they reasoned that proteases might be a secreted product of tumor cell metabolism. Reich's group has shown that in a number of instances transformed

Table 1

Summary of Effects of Proteases on Composition of
Cell Surface Proteins, DNA Synthesis, Cell
Proliferation and Aggregation of Cells

Enzyme treatment (μg/ml)	Relative amount of ^{131}I protein		Relative DNA synthesis	Relative cell number	Cell Aggregation
	250K	205K	12 hr	36 hr	
None (control)	100	100	1	1	
Serum (5%)	100	100	19.4	2.36	
Thrombin (2.5)	100	0	23.5	2.25	+
α-Protease (5.0)	100	0	1.3	1	+
Thermolysin (5.0)	0	0	2.2	1.09	−
Bromelin (5.0)	100	0	14.1	2.10	+
Trypsin (0.1)	0	0	16.4	2.13	−
Papain (1.0)	0	0	1.6	1.05	−
Elastase (5.0)	0	0	11.8	1.69	−

or tumor cells secrete a factor, which after reaction with serum, yields a proteolytic enzyme capable of catalyzing the hydrolysis of fibrin films. The cell factor was identified by them as a plasminogen activator and the serum factor as plasminogen. The active agent concerned with fibrinolysis was, of course, plasmin. A similar finding was made by Goldberg (30) by use of casein as substrate instead of fibrin. In view of these observations, particularly since plasmin has been demonstrated to be mitogenic (30), it was tempting to many to attribute a more profound and specialized role to plasmin in the maintenance and growth of the tumor cell. According to this earlier model, plasmin as a protease was postulated to stimulate tumor cells giving them a preferential advantage in cell division and growth. The development of this simple model required 1) the demonstration of its role as a mitogen particularly of transformed cells and 2) the obligatory participation of plasmin in the transformation process.

We have attempted to contribute further information about the second of these two requirements. For these investigations we were fortunate to have available through the generosity of Dr. David F. Waugh of this department, highly purified reagents in the form of fibrinogen and thrombin.* A simplified assay for fibrinolysis was then developed in which ^{125}I-iodinated fibrinogen was reacted with thrombin to form a clot on a petri dish in which either normal or Rous sarcoma virus-chick embryo fibroblasts had been seeded and incubated in Dulbecco modified Eagle's medium in the absence of serum (42). The fibrin clot formed by this method is thus not cross-linked since Factor XIII or transglutaminase, a normal constituent of serum, had been excluded from the incubation. The conditions of our assay, therefore, differ importantly from those used by Reich's group. These investigators coated ^{125}I-iodinated fibrinogen on a petri dish and then formed an insoluble clot by addition of serum. Since their assay required the use of serum, the fibrin clot was cross-linked. This form of fibrin is enzymatically hydrolyzed with somewhat more difficulty than is the uncross-linked variety.

Since our principal objective was to test the role of plasmin in the maintenance of the transformed phenotype, all of the reagents, including the serum used for growth of either the normal or transformed chick embryo fibroblasts, were exhaustively treated with lysine-Sepharose beads to remove plasminogen (43). The digestion of ^{125}I-labeled fibrin was then measured by release of soluble radioactive peptides into the medium.

As a verification that our reagents were plasminogen-

---

* Fibrinolysis was not observed when an attempt was made in another laboratory to repeat our experiments on RSV-transformed chick cells with an impure, commercial preparation of fibrinogen to which plasminogen had not been added. However, identical results as described by Chen and Buchanan (42) could be obtained by them with fibrinogen supplied by this laboratory. Since there have been other contradictions in the results obtained by several laboratories working in the field of fibrinolysis, we believe that many discrepancies can be accounted for by variations in the purity of the fibrinogen and thrombin prepations used. We, therefore, recommend that only reagents of highest purity be employed in order to avoid inhibition by impurities not only of proteases in general but also of plasminogen activator and plasmin, themselves.

and plasmin-free we could demonstrate that clots formed by the procedure described above remained solid for over 4 days when incubated at 37° in the presence of added urokinase. When incubated with normal fibroblasts, fibrin clots did not dissolve at a measurable rate for at least 4 days. However, in the presence of RSV-transformed chick embryo fibroblasts lysis of the fibrin clot and the release of radioactive peptides occurred within the first three hours. The rate of proteolysis was roughly related to the concentration of cells seeded on the plate (42).

The participation of a viral gene product in the lytic process was demonstrated by use of Kawai and Hanafusa's temperature sensitive mutant (ts68) of the Schmidt-Ruppin strain of Rous sarcoma virus, subgroup A (44). The characteristics of the transformed phenotype can be observed at the permissive temperature of 36° but not at the non-permissive temperature, 41°. As shown in Fig. 3 a shift in the temperature from 41° to 36° and vice versa results in the increase or decrease in the production of fibrinolysins, respectively, within a period of 2 to 3 hrs. The change is complete by the 17th to 20th hr in either direction (42).

The important feature of the experiments with either wild type or temperature-sensitive virus is that in both cases the fibrinolysis occurred in the absence of plasminogen. Thus, the active agent or enzyme involved in fibrinolysis is not produced as a result of the formation of plasmin from plasminogen in a reaction catalyzed by plasminogen activator secreted by the transformed cells. The enzyme responsible for plasminogen-independent fibrinolysis could be the same protease, which under other conditions activates plasminogen to plasmin, or alternatively, an entirely different enzyme, which is also secreted by transformed cells. These experiments in no way contradict the fundamental experimental observation of the Rockefeller group that plasminogen activator is formed and secreted by the RSV-transformed chick embryo fibroblast. These experiments do, however, show that other proteolytic activities can be demonstrated under these experimental conditions, and attempt to evaluate the role of plasmin in the transformation process.

The latter question was further explored by measurement of the role of plasmin in the expression of several of the characteristic features of the transformed phenotype. Specifically, growth rate and final density, uptake of

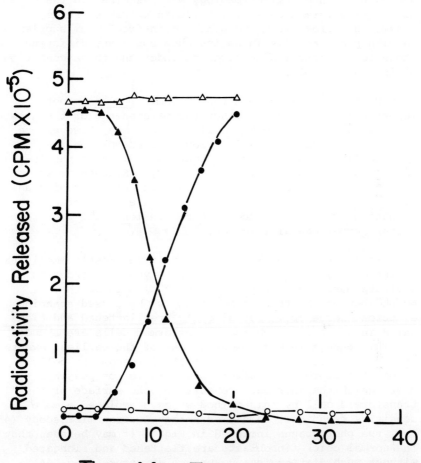

Fig. 3. Fibrinolytic activity of Ts68-infected chick cells during temperature shift (42) 36° → 36° (△--△); 36° → 41° (▲--▲); 41° → 41° (O--O); 41° → 36° (●--●). The individual points represent assay of individual plates of cells whose medium was replaced with Dulbecco-modified Eagle's medium at the indicated time and to which was added ^{125}I fibrinogen and thrombin. The release of radioactive peptides over a three-hour period was measured.

deoxyglucose and cell morphology were examined in the case of RSV-transformed cells incubated in untreated serum, plasminogen-free serum and plasminogen-free serum supplemented with plasminogen. In no instance could any difference in these four properties be observed under the three conditions of incubation (42).

We therefore conclude that plasmin per se is not importantly involved in either the maintenance or the growth stimulation of transformed cells. There may, however, be other aspects of plasmin function, for example on the permeability of blood vessels that could lead to the invasiveness of cancer cells in the body. Neither do we intend to imply that any of the secreted products of transformed cells have been directly implicated in a transformation cycle. This important conclusion still requires further experimental evidence.

So far the morphology of cells is the most commonly used criterion of the transformation phenotype. In order to evaluate more decisively the role of plasmin on the morphology of the transformed cell, we have used scanning electron microscopy. Normal cells are flattened and elongated in appearance, whereas transformed cells are rounded up. This change in morphology is one of the earliest events that can be observed with the light microscope as cells become transformed. However, since it is not possible to distinguish the intricate features of the surface of the transformed cell by this method we have employed a more sensitive technique, namely scanning electron microscopy (45). From the photographs included in Fig. 4 it may be seen that the normal chick fibroblasts are flattened and elongated without much distinguishing surface structure (Fig. 4A). However, in the case of chick cells infected with the temperature-sensitive viral mutant, very characteristic structures resembling flowers or ruffles appear on the cell surface as the temperature is shifted from the nonpermissive to permissive range (Fig. 4B, C and D). A highly magnified surface flower seen in Fig. 5 reveals the interconnections of the ruffle folds.

These ruffles appear when incubation is carried out in a plasminogen-free medium and in the presence of inhibitors of protein synthesis. Their first appearance occurs within 30 min after the temperature shift and is essentially complete for those cells that undergo transformation by the

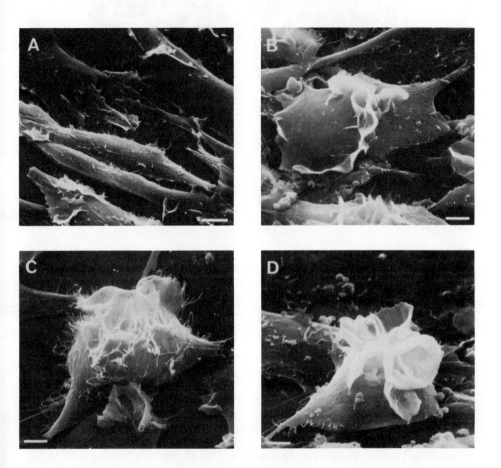

Fig. 4. Scanning electron micrographs of Ts68-infected chick embryo fibroblasts. Cells were maintained at 41° for 48 hr and fixed either before or at various times after a shift to the permissive temperature. (A) Surface of cells maintained at 41° bar = 5 μm. (B) Cells fixed 1 hr after shift from 41° to 36°; bar = 3 μm. (C and D) Cells fixed 2 hr after shift from 41° to 36°; bar = 3 μm.

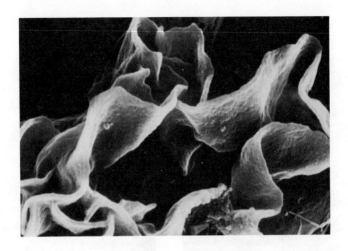

Fig. 5. A high magnification of a surface ruffle formed in the presence of an inhibitor of protein synthesis. Ts68-infected chick embryo fibroblasts were maintained at 41° for 48 hr. The temperature was then shifted to 36° with the addition of cycloheximide at a concentration of 5 μg/ml. The cells were fixed after 2 hr of incubation for scanning electron microscopy.

end of the second hour. The appearance of surface ruffles is thus one of the earliest events observed to occur during the transformation process. The secretion of fibrinolysins occurs between the eighth and twentieth hour (42) and the intracellular level of plasminogen activator increases after 4 hr (46). The loss of the LETS or 250K protein occurs after the eighth hour (20). The increase in glucose uptake takes place sometime after the third or fourth hour (44).

The ability of surface ruffles to be formed in the presence of inhibitors of protein synthesis added at time of temperature shift indicates the viral gene product responsible for ruffling has been produced during incubation at the nonpermissive temperature but does not express its activity until the temperature is lowered.

At present we feel that surface ruffles can be

designated only as markers of the transformation process. We do not know, for example, whether they are a causitive agent or rather a product of some reaction catalyzed by a recently activated viral-specific cell component such as an intracellular protease.

DISCUSSION

Use of proteases as mitogens to probe the complexity of reactions concerned with the stimulation of resting or dormant cells to proliferation has yielded some information about this process. The two blood proteases of particular concern in this study have been thrombin and plasmin. Thrombin, a highly specific protease, stimulates DNA synthesis and brings about an increase in cell number when added as a single macromolecular reagent to secondary chick embryo and murine fibroblasts. It does not, however, stimulate RSV-transformed cells but can act in concert with other mitogens to promote the growth of certain line cells.

Thrombin has been particularly useful in studies attempting to correlate changes in the composition of cell surface proteins with certain cellular properties. Two cell surface proteins, one of 250,000 and the other of 205,000 daltons, have received special attention as possible candidates concerned with the control of growth and cellular aggregation. Neither protein can be directly or simply related to growth control. The smaller but not the larger protein is removed during reaction of thrombin with cell surface components. The role of thrombin in stimulating cell growth must therefore be attributed to a further function of this enzyme after the initial action has occurred.

A striking pattern of cell aggregation is observed with thrombin-treated chick fibroblasts. The aggregation of cells after treatment with thrombin is ascribed first to removal of the 205K protein with a subsequent morphological change characterized by reduction of both cytological spreading and cell-cell interactions. Then, possibly because of greater exposure of the 250K protein on the cell surface, there is enhanced intercellular adhesion and aggregation of cells from which the 250K protein was not cleaved by protease action.

These studies have cast some light on the chemical alterations of the cell surface, but have only begun to contribute to knowledge of the intracellular reactions of fibroblasts stimulated by the mitogenic proteases. One interesting report has been the finding that brief treatment of methylcholanthrene-transformed mouse BalB/3T3 cells with thrombin results in the production of prostaglandin E_2 and $F_{2\alpha}$ in particular the former (47). This observation has been coupled with the recent observation (35) that prostaglandin $F_{2\alpha}$ can extend the mitogenic effect of 5% serum in Swiss 3T3 mouse cells. We have confirmed this latter observation, but on the basis of our present experiments are not able to ascribe the several effects of thrombin solely to its stimulation of prostaglandins.

In regard to the role of plasmin in the maintenance and mitogenic stimulation of RSV-transformed chick embryo fibroblasts, we have provided evidence that fibrinolysis can occur in the absence of plasmin. A protease or proteases are secreted by these cells that are capable of acting directly on a fibrin clot to cause its dissolution. A similar observation has been made by Plow and Edgington (48) in which leucocytes were found to possess plasminogen-independent fibrinolytic activity in addition to a plasminogen-dependent system. Dr. Yunis will report in this Symposium his own findings with a rat mammary carcinoma.

This in no way lessens the importance of the finding that plasminogen activator is an important secretion of the transformed cell. Our studies do not confirm the putative role of plasmin as an agent causing the autostimulation of these cells and thus conferring on them a growth advantage.

The appearance of surface ruffles on cells as they become transformed by manipulation of the incubation temperature of cells infected with a temperature-sensitive Rous sarcoma virus is one of the earlier markers of this process. The lack of requirement for protein synthesis for this development signifies that the viral gene product has already been formed but is incapable of expressing itself unless incubation occurs at the permissive temperature. The intricate biochemical reactions leading to this step in viral transformation still remain to be determined.

REFERENCES

1. G. Todaro, Y. Matsuya, S. Bloom, A. Robbins and H. Green, in: Growth Regulating Substances for Animal Cells in Culture, Wistar Inst. Symp. Monogr. no. 7, 1967, p. 87.

2. R.W. Pierson, Jr. and H.M. Temin, J. Cell Physiol. 79 (1972) 319.

3. R.W. Holley and J.A. Kiernan, Proc. Nat. Acad. Sci. USA 60, (1968) 300.

4. H.A. Armelin, K. Nishikawa and G.H. Sato, in: Cold Spring Harbor Conferences on Cell Proliferation, vol 1, Control of Proliferation in Animal Cells, eds. B. Clarkson and R. Baserga (Cold Spring Harbor Laboratory, 1974) p.97.

5. H.M. Temin, J . Cell. Physiol. 69 (1967) 377.

6. W.H. Daughaday, K. Hall, M.S. Raben, W.D. Salmon, J.L. Van den Brande and J.J. Van Wyk, Nature 235 (1972) 107.

7. D. Gospodarowicz, Nature 249 (1974) 123.

8. J.M. Taylor, W.M. Mitchell and S. Cohen, J. Biol. Chem. 247 (1972) 5928.

9. H.W. Fisher, T.T. Puck and G. Sato, Proc. Nat. Acad. Sci. 44 (1958) 4.

10. G.M. Healy and R.C. Parker, J. Cell Biol. 30 (1966) 539.

11. A.Vaheri, E. Ruoslahti, M. Sarvas, M. Nurminen, J. Exp. Med. 138 (1973) 1356.

12. C.R. Thrash and D. D. Cunningham, Nature 242 (1973) 399.

13. B.M. Sefton and H. Rubin, Nature 227 (1970) 843.

14. M.M. Burger, Nature 227 (1970) 170.

15. L. Ossowski, J.P. Quigley, G.M. Kellerman and E. Reich, J. Exp. Med. 138 (1973) 1056.

16. M.M. Burger, in: Growth Control in Cell Cultures, Ciba Foundation eds. G.E.W. Wolstenholme and J. Knight (Churchill Livingstone, London (1971)) p.45.

17. R.O. Hynes, Proc. Nat. Acad. Sci. USA 70 (1973) 3170.

18. R.O. Hynes, and K.C. Humphryes, J. Cell Biol. 62 (1974) 438.

19. R.O. Hynes, Cell 1 (1974) 147.

20. P.W. Robbins, G.G. Wickus, P.E. Branton, B.J. Gaffney, C.B. Hirschberg, P.Fuchs and P.M. Blumberg in: Cold Spring Harbor Symposium on Quantitative Biology, vol.39 (Cold Spring Harbor Laboratory, Cold Spring Harbor, N.Y., 1974) p. 1173.

21. N.M. Hogg, Proc. Nat. Acad. Sci. USA 71 (1974) 489.

22. K.R. Stone, R.E. Smith and W.K. Joklik, Virology 58 (1974) 86.

23. C.G. Gahmberg and S. Hakomori, Biochem. Biophys. Res. Commun. 59 (1974) 283.

24. E. Ruoslahti, A. Vaheri, P. Kuusela and E. Linder, Biochim. Biophys. Acta. 322 (1973) 352.

25. E. Ruoslahti and A. Vaheri, Nature 248 (1974) 789.

26. J.C. Unkeless, A. Tobias, L. Ossowski, J. Quigley, D.B. Rifkin and E. Reich, J. Exp. Med. 137 (1973) 85.

27. J.C. Unkeless, K. Danø, G.M. Kellerman and E. Reich, J. Biol. Chem. 249 (1974) 4295.

28. J.P. Quigley, L. Ossowski and E. Reich, J. Biol. Chem. 249 (1974) 4306.

29. L. Ossowski, J.P. Quigley, and E. Reich, J. Biol. Chem. 249 (1974) 4312.

30. A.R. Goldberg, Cell, 2 (1974) 95.

31. L.B. Chen, N.N.H. Teng and J.M. Buchanan, unpublished results.

32. S. Magnusson, in: The Enzymes, Third Ed. Vol III, ed. P.D. Boyer (Academic Press, New York, 1971) p. 277.

33. L. Muszbek and K. Laki, Proc. Nat. Acad. Sci. USA 71 (1974) 2208.

34. L.B. Chen and J.M. Buchanan, Proc. Nat. Acad. Sci. USA 72 (1975) 131.

35. L.J. DeAsua, D. Clingan and P.S. Rudland, Proc. Nat. Acad. Sci. USA 72 (1975) 2724.

36. K. J. Lembach, Proc. Nat. Acad. Sci. USA 73 (1976) in press.

37. N.N.H. Teng and L.B. Chen, Proc. Nat. Acad. Sci. USA 72 (1975) 413.

38. N.N.H. Teng and L.B. Chen, Nature (1976) in press.

39. B.R. Zetter, L.B. Chen and J.M. Buchanan, Cell (1976) in press.

40. K.M. Yamada, S.S. Yamada and I. Pastan, Proc. Nat. Acad. Sci. USA 72 (1975) 3158.

41. A. Fischer, Arch. Entwicklungsmech. Org. (Wilhelm Roux) 104 (1925) 210.

42. L.B. Chen and J.M. Buchanan, Proc. Nat. Acad. Sci. USA 72 (1975) 1132.

43. D.G. Deutsch and E.T. Mertz, Science 170 (1970) 1095.

44. S. Kawai and H. Hanafusa, Virology 46 (1971) 470.

45. V.R. Ambros, L.B. Chen and J.M. Buchanan, Proc. Nat. Acad. Sci. USA 72 (1975) 3144.

46. D.B. Rifkin, L.P. Beal and E. Reich in: Proteases and Biological Control, eds. E. Reich, D.B. Rifkin, and E. Shaw (Cold Spring Harbor Laboratory, Cold Spring Harbor, N.Y. 1975) p. 841.

47. S.C. Hong, R. Polsky-Cynkin and L. Levine, J . Biol Chem. (1976) in press.

48. E.F. Plow and T.S. Edgington, J. Clin. Invest. 56 (1975) 30.

ACKNOWLEDGMENTS

The work cited from this laboratory was supported by a grant-in-aid from the National Science Foundation (BMS17669), the American Cancer Society (BC206) and the National

Cancer Institute (CA02015). BRZ was a postdoctoral fellow of the Damon Runyon-Walter Winchell Cancer Fund and LBC a predoctoral fellow of the Johnson and Johnson Co. Epidermal growth factor was kindly supplied by Dr. T-T Sun of this Department. The 3T3 cells infected with Avian sarcoma virus B-77 were kindly supplied by Dr. Thomas Benjamin.

Discussion

S. Magnusson, University of Århus: I think these results with thrombin as a mitogenic agent are extremely stimulating for us. We have been working on the structure and I think it would be very nice if one could possibly try to find out which part of the structure of thrombin is important in this context, and I have a few questions in relation to that. The first is that one of the platelet groups in this country has been finding that the DFP inhibited thrombin does not induce platelet aggregation, but they find that if they incubate platelets with DFP inhibited thrombin they can no longer induce aggregation by adding active thrombin indicating that there may be a binding site on the thrombin molecule which is in a different position from that of the active site. I wonder if you have any results on that with your cell systems.

J. Buchanan, Massachusetts Institute of Technology: Yes, we have tried to obtain stimulation of chick embryo fibroblasts with DFP-inhibited thrombin without success. Therefore, the unblocked active site of thrombin is necessary for its mitogenicity. Perhaps more to the point of your question, we have compared the relative mitogenicities of alpha, beta and gamma thrombins in which there is a reduction of the clotting activity of the latter two materials without loss of esterase activity. Our attempts to show that the mitogenicity of thrombin is due to its esterase activity have so far not been conclusive. Our main problem is that we have had trouble in obtaining thrombin derivatives completely free of α thrombin. If you have samples of that, we would certainly like to try it.

S. Magnusson: Maybe we can discuss that later. We do not really have any esterase thrombin. But the second question I would like to make is, if one tries to sort of condense the differences between thrombin and the other serum proteases in primary structure terms, there seem to be three main differences in the B chain: one is that thrombin has this carbohydrate on the insertion at position 65A, the second is the homology with the angiotensin structure which may be another binding site, and the third is the 168/182 disulfide bridge structure which is homologous with the leuteinizing hormone. I wonder if you have done any competition experiments with thrombin using one of the hormones to block the mitogenic activity of thrombin.

    J. Buchanan: We have not yet carried out experiments of the type you suggest.

THE SECRETION OF FIBRINOLYSIN BY CULTURED MAMMALIAN TUMOR CELLS*

M. C. WU, D. R. SCHULTZ AND A. A. YUNIS**
The Howard Hughes Medical Institute, and Departments
of Medicine and Biochemistry, University of Miami
School of Medicine, Miami, Florida 33152

INTRODUCTION

In the course of investigations by one of the authors (A.Y.) on cultured rat ovarian tumor cells (1, 2) in the laboratory of Dr. G. Sato (summer of 1972), gross fibrinolysis was observed around cells grown on fibrin plates. Microscopic observation of this process showed the attachment of cells to fibrin, and in 1 to 2 days, the cells were surrounded by a halo of fibrinolysis (3). When a cell lysed its way through the fibrin, it attached to the bottom of the plate and formed a colony. In 6-8 days, the plate was filled with a variable number of holes, corresponding to the number of colonies. We subsequently showed that this "fibrinolytic activity" was secreted by these cells and could be isolated from conditioned serum free medium (3). At about the same time, investigators at Rockfeller University described increased fibrinolytic activity associated with viral-induced cell transformation (4, 5).

The important questions arising from these observations concerned the nature of this fibrinolytic activity and its relationship to the neoplastic process.

Since our initial observations on rat ovarian tumor cells a number of mammalian tumor cell lines have been established in our laboratory, all of which secrete fibrinolytic

* This work was supported by Howard Hughes Medical Inst.

** Howard Hughes Investigator.

activity. These include rat prostate carcinoma, R2426 breast carcinoma and sarcoma, and human breast carcinoma, melanoma, and pancreatic carcinoma. In this presentation we describe the purification and characterization of "fibrinolysin" from rat breast carcinoma R2426. This particular cell line was chosen because it has retained its tumorogenic properties when injected into recipient rats and thus offers a potential experimental model for investigating the possible role of fibrinolysis in neoplasia both in vitro and in vivo.

EXPERIMENTAL AND RESULTS

Cell Cultivation and Assay of Fibrinolysis

The rat breast carcinoma R2426 cell line was established as described previously (6). Cells were grown in Dulbecco's Modified Eagle's Medium with 10 per cent fetal calf serum and 2.5 per cent horse serum. For the isolation and purification of fibrinolysin, culture plates with cells grown from 4 to 7 days were rinsed thoroughly with Hank's balanced salt solution, serum-free medium was added, the plates were incubated for an additional 48 hr, and the conditioned medium was collected.

Three methods were used to assay fibrinolytic activity: (a) the clearing of translucent fibrin-agar plates, (b) the hydrolysis of ^{125}I-fibrinogen, (c) the hydrolysis of N-p-tosyl-L-arginine methyl ester (TAME).

Nature of Fibrinolytic Activity Secreted by R2426 Cells.

The fibrinolytic activity observed in R2426 tumor cell cultures could represent an activator of a pro-fibrinolysin, (e.g. plasminogen activator), or a direct fibrinolysin similar or identical to plasmin. We have established the presence of both plasminogen activator and direct fibrinolysin activities in the conditioned serum free medium (6). Figure 1A shows the results of the ^{125}I-fibrinogen hydrolysis assay. Plasminogen alone had some activity, probably because of slight contamination with plasmin. However, the conditioned medium had considerable direct fibrinolytic activity. The addition of conditioned medium to plasminogen caused an increase in activity in excess of the sum of both individual activities, indicating the presence of an activator of a zymogen in addition to direct fibrinolytic activity. Also shown here is the activation of plasminogen by urokinase as a positive control. Similar results were obtained from the TAME assay as illustrated in Figure 1B.

Characteristics of the Direct Fibrinolysin

The direct fibrinolysin was partially purified from conditioned serum free medium by DEAE-Cellulose and Sephadex G-200 gel filtration chromatography (6). Fractions with fibrinolytic activity from the DEAE-Cellulose chromatographic separation were pooled, concentrated, and applied to Sephadex G-200. The separation of activator and direct fibrinolytic activities on Sephadex G-200 is shown in Fig. 2. Two different assays were used for these experiments: The hydrolysis of ^{125}I-fibrinogen and the clearing of fibrin-agar plates. Assay of the fractions by the hydrolysis of plasminogen-free ^{125}I-fibrinogen yielded a single peak of direct fibrinolytic activity. When the fractions were preincubated with plasminogen and then assayed, a second peak appeared representing plasmin activity generated by the activation of plasminogen; thus, the direct fibrinolysin and the plasminogen activator were clearly separable. Finally, when the fractions were assayed in unheated fibrin-agar plates, a broad peak overlapping the other two peaks was obtained, indicating the combined action of a direct fibrinolysin and a plasminogen activator.

The purified direct fibrinolysin had an estimated molecular weight of 110,000 daltons, it was heat stable (60°, 12 hr), and it showed direct hydrolytic activity by all three methods of assay. Like plasmin, R2426 fibrinolysin was inhibited by $\mathcal{E}$-amino caproic acid and soybean trypsin inhibitor, and required Ca^{++} for the hydrolysis of TAME. However, it differed in several respects from plasmin including a higher molecular weight, a faster electrophoretic mobility in polyacrylamide gel and a distinctly different fibrin degradation product profile as determined in polyacrylamide gel electrophoresis.

These experiments established the presence of a direct fibrinolysin in serum free conditioned medium from R2426 rat breast carcinoma cells. The possibility that this represented contamination with plasmin from fetal calf serum and/or horse serum was unlikely since the cultured cells were washed thoroughly before the addition of serum free medium. However, to further exclude this possibility, R2426 cells were grown in plasminogen-free growth medium for several generations, the medium was prepared as above, and assayed for activity. Direct fibrinolytic activity was again demonstrable by all three assay methods, leaving little doubt that the enzyme originated from the cells and not from the calf or horse serum components. Further evidence supporting this conclusion was obtained from immunologic studies (7).

Preparation and Action of Anti-Fibrinolysin Antibody.

For the purpose of immunization, the enzyme preparation from Sephadex G-200 chromatography was further purified by gel electrophoresis and the corresponding slices containing the activity were pooled, homogenized with complete Freund's adjuvant, and injected intramuscularly into a goat's hind legs. The animal received 3 injections over the course of three months and was bled 5 days after the last injection. The antibody thus generated was partially purified and characterized as a 7S heat stable (56°C, 30 min.) nonprecipitating type which neutralized fibrinolytic activity as assayed by the hydrolysis of I^{125}-fibrinogen and TAME. The effect of increasing antibody concentration on fibrinolysis in unheated fibrin-agar plates is shown in Fig. 3; a correspondingly increasing degree of inhibition was observed.

Immunologic Relationship of R2426 Fibrinolysin to Other Fibrinolysins.

Anti-R2426 fibrinolysin antibody cross reacted with fibrinolysins secreted by cultured human melanoma and human breast carcinoma cells, and to some preparations of partially purified human plasmin. However, inhibition of human plasmin activity was greatly reduced after absorption of goat anti-R2426 fibrinolysin IgG fraction with human plasminogen, but the fraction remained completely inhibitory for the R2426 fibrinolysin. Furthermore, the antibody had no inhibitory effect on either horse or fetal calf plasmin. Thus, the fibrinolysin secreted by R2426 cells was immunologically distinct from human, horse and calf plasmin.

Characteristics of R2426 Plasminogen-Activator

The plasminogen activator from R2426 serum-free conditioned medium was purified 2000-fold with an overall recovery of ca. 40 per cent. The purification procedure included affinity chromatography on Sepharose-arginine methyl ester followed by filtration on Sephadex G-200 (Fig. 4). The purified activator was inhibited by diisopropylfluorophosphate (DFP), a property similar to that of activator from virus transformed cells (8, 9) and other tumor cell lines (10, 11). In comparative experiments, both urokinase and R2426 plasminogen activator were labelled with ^{3}H-DFP and examined by SDS polyacrylamide gel electrophoresis (Fig. 5A and B). The ^{3}H-DFP radioactivity coincided with fibrinolytic activity for both the R2426 activator and urokinase, and the profile indicated that both factors have similar molecular weights (ca. 55,000 D). The M.W. is greater than the plasminogen activator from Rous sarcoma virus-transformed chick embryo fibroblasts (39,000 D) (8). In addition, the commercial urokinase (Calbiochem) and the crude serum free conditioned medium from R2426 cells showed two distinct activities in SDS-gel.

Lesuk (12) suggested that the two distinct activities observed with urokinase resulted from proteolytic action of pepsin in the urine. Whether the two R2426 plasminogen activator activities in crude serum free conditioned medium represent a similar phenomenon or the existence of two isoenzymes is uncertain at present.

The mechanism of activation of plasminogen by urokinase has been well studied (13, 14). Plasminogen, which is a single peptide of M.W. 87,000 daltons, is converted upon activation by urokinase to two smaller peptides of 55,000 and 26,000 daltons linked by one disulfide bond. As can be seen in Fig. 6, the activation of plasminogen by R2426 activator follows a similar mechanism.

CONCLUDING REMARKS

We have shown in these studies that cultured rat breast carcinoma R2426 cells secrete both a plasminogen activator and a direct fibrinolysin. The plasminogen activator is similar to urokinase in molecular weight and in its mode of activation of plasminogen. The direct fibrinolysin hydrolyses fibrinogen or fibrin directly, but appears to be distinct from plasmin as shown both by biochemical and immunological studies. It may be similar to the plasminogen-independent protease activity from Rous sarcoma virus-transformed chick embryo fibroblasts recently described by Chen and Buchanan (15).

Our studies and those of others (4, 5) suggest strongly that the secretion of fibrinolytic activity is a property of neoplastic cells. We have not been able to demonstrate any fibrinolytic activity in established cultures of various mammalian fibroblast lines. In our attempts to establish various tumor cell lines, we have observed that fibrinolytic activity is present in conditioned medium from primary cultures but disappears as the tumor cells become replaced by fibroblasts. On the other hand, when the tumor cell colonies predominate and the fibroblasts disappear with subsequent transfers, fibrinolytic activity reappears and is maintained as the tumor cell line becomes established. Thus, the reappearance and/or persistence of fibrinolytic activity in the medium is a good indication that a given tumor cell line is established.

The role of the increased fibrinolytic activity in neoplasia remains uncertain. It could serve an important function in tumor cell adaptation and invasiveness, since fibrin deposition may be one mechanism of host defense against tumor spread. Both in vitro and in vivo models can be devised to study this aspect of tumor-host interaction. An in vitro model that appears to be worthy of investigating is the growth of tumor cells in fibrin plates. When R2426 cells are grown on fibrin plates the cells become attached immediately to the fibrin and begin lysing their way to the bottom of the plate where they form colonies. It is of interest that this "lytic" invasion can be retarded or prevented by incorporating anti R2426 fibrinolysin antibody into the fibrin (Fig. 7). The physiologic significance of these observations awaits further studies.

REFERENCES

(1) G. H. Sato, in Molecular Genetics and Develop. Biol., M. Sussman Editor. Prentice Hall, N.J. pp. 475 (1971).

(2) J. L. Clark, K. K. Jones, D. Gospodarowics, and G. H. Sato, Nature New Biol., 236 (1972) 180.

(3) A. A. Yunis, D. R. Schultz and G. H. Sato, Biochem. Biophys. Res. Comm., 52 (1973) 1003.

(4) J. C. Unkeless, A. Tobia, L. Ossowski, J. P. Quigley, D. B. Rifkin and E. Reich. J. Exp. Med., 137 (1973) 85.

(5) L. Ossowski, J. C. Unkeless, A. Tobia, J. P. Quigley, D. B. Rifkin, and E. Reich, J. Exp. Med., 137 (1973) 112.

(6) M.-C. Wu, D. R. Schultz, G. K. Arimura, M. A. Gross and A. A. Yunis. Exptl. Cell Res., 96 (1975) 37.

(7) D. R. Schultz, M.-C. Wu, and A. A. Yunis, Exptl Cell Res, 96 (1975) 47.

(8) J. C. Unkeless, K. Dano, G. M. Kellerman, and E. Reich, J. Biol. Chem., 249 (1974) 4295.

(9) J. K. Christman and G. Acs, Biochim. Biophys. Acta 340 (1974) 339.

(10) J. R. Wachsman and J. L. Biedler, Exptl Cell Res., 86 (1974) 264.

(11) D. B. Rifkin, J. N. Loeb, G. Moore, and E. Reich, J. Exp. Med., 139 (1974) 1317.

(12) A. Lesuk, L. Termineillo, J. H. Traver, and J. L. Groff, Thromb. Diath. Haemorrh., 18 (1967) 293.

(13) L. Summaria, L. Arzadon, P. Bernabe, and K. C. Robbins. J. Biol. Chem., 250 (1975) 3988.

(14) P. J. Walther, H. M. Steinman, R. L. Hill, and P. A. McKee. J. Biol. Chem. 249 (1974) 1173.

(15) L. B. Chen and J. M. Buchanan, Proc. Nat. Acad. Sci. USA, 72 (1975) 1132.

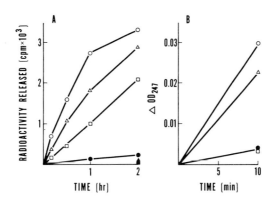

Figure 1 (A) [^{125}I] fibrinogen hydrolysis assay: The incubation mixture, in a final volume of 0.3 ml, contained 0.15 M NaCl, 0.01 M sodium phosphate pH 7.5, [^{125}I] fibrinogen (3 x 10^5 cpm), and enzyme samples as indicated below. Enzyme samples were preincubated at room temperature for 30 min before the addition of [^{125}I] fibrinogen, then at 37°C, and a 50 µl aliquot was removed at different time intervals and analysed. Control (■–■); human plasminogen (10 µg) (●–●); urokinase (2.5 U) (▲–▲; R2426 rat fibrinolysin (100 µl, 0.15 mg/ml) (□–□); R2426 fibrinolysin (100 µl + plasminogen (10 µg) (○–○); urokinase (2.5 U) + plasminogen (10 µg) (△–△); (B) TAME hydrolysis assay: The reaction mixture in a final volume of 1.0 ml contained 0.04 M Tris-HCl, pH 8.1, 0.01 M CaCl$_2$, 0.001 M TAME, and the enzyme samples. The enzyme samples were preincubated at 25°C for 30 min. Reaction rate was measured by the increase in optical density at 247 nm as described in the methods. Control (■–■); human plasminogen (10 µg) (●–●); urokinase (2.5 U) (▲–▲); R2426 rat fibrinolysin (100 µl, 0.15 mg/ml) (□–□); R2426 fibrinolysin (100 µl) + plasminogen (10 µg) (○–○); urokinase (2.5 U) + plasminogen (10 µg) (△–△). (From Wu et al, Exptl Cell Res. 96: 37, 1975, by permission from the publishers).

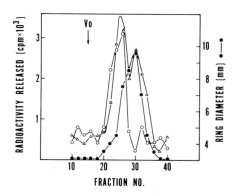

Figure 2   Separation of rat fibrinolysin from plasminogen activator activity on Sephadex G-200. Concentrated P II (0.5 ml, 3 mg/ml) was rechromatographed on a 1.5 x 60 cm Sephadex G-200 column equilibrated with 0.3 M NaCl, 0.01 M sodium phosphate buffer, pH 7.5. Fractions of 1.5 ml were collected and assayed for direct fibrinolysin and plasminogen activator activity by the hydrolysis of $[^{125}I]$ fibrinogen and by the unheated fibrin-agar plate technique. For the direct fibrinolytic activity assay (O–O) 100 μl aliquots were removed from each fraction and added to plasminogen-free $[^{125}I]$ fibrinogen (5 x 10^4 cpm) (assay procedure described under fig. 5). A single activity peak was observed. Aliquots were also assayed for plasminogen activator by incubation with plasminogen followed by assay of fibrinolytic activity by the $[^{125}I]$ fibrinogen method (△–△). Two peaks were observed, one representing the first peak of direct fibrinolytic activity and the second representing the plasmin generated by the activation of plasminogen (activator activity). The assay using the unheated fibrin-agar plate technique (●–●) yielded a broad peak overlapping both activity peaks thus representing both direct fibrinolysis and activation of plasminogen which is a contaminant of the fibrinogen used in the plate assay. (From Wu et al., Exptl. Cell Res. 96: 37, 1975, by permission from the publishers).

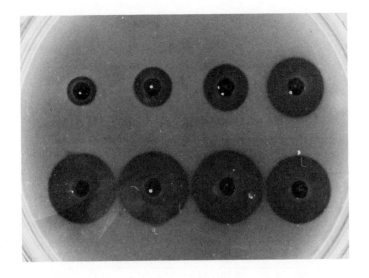

Figure 3 Inhibitory effects of anti-rat R2426 fibrin olysin antibody on rat R2426 fibrinolysin activity as assayed in an unheated fibrin-agar plate: (Top) Serial dilutions of goat anti-rat R2426 fibrinolysin and the R2426 fibrinolysin (v/v) in serum-free conditioned medium. (Bottom) same as top except an equivalent normal goat serum fraction was substituted for the immune fraction. (From Schultz, Wu and Yunis, Exptl Cell Res. 96: 47, 1975, by permission from the publishers).

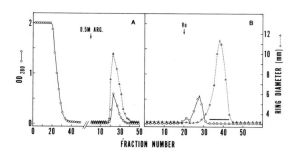

Figure 4   Purification of Plasminogen Activator

(A) Sepharose - Arginine Methyl Ester Affinity Chromatography: Conditioned serum free medium (1000 ml) was concentrated to 50 ml. Triton X-100 (1%) and Sepharose-Arginine Methyl ester (50 ml) were added to the concentrate and the mixture was stirred gently overnight and then passed through the column. After washing with 0.1 M Tris-HDl pH 8.0 and 0.5 M potassium phosphate buffer pH 7.4 containing 3 mM EDTA (20 column volumes each), the activator was eluted with 0.5 M Arginine in 0.1 M phosphate, 3 mM EDTA, buffer pH 7.4.

(B) Sephadex G-200 Gel Filtration: Activator from step (A) was pooled, concentrated and applied to a Sephadex G-200 column pre-equilibrated with 0.3 M NaCl in 0.005 M phosphate buffer pH 7.4.

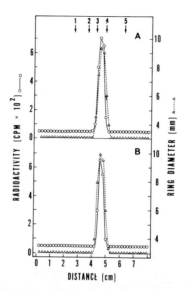

Figure 5 SDS gel electrophoresis of R2426 Plasminogen Activator (A) and Urokinase (B): Purified R2426 plasminogen activator and urokinase which were either labelled with ^{3}H diisopropyl fluorophosphate (DFP) or unlabelled were run in 7.5% gels. Gel slices (2 mm slices) were assayed for fibrinolytic activity on fibrin-agar plates. The radioactivity of corresponding slices was determined in Triton X-100 scintillation fluid. Position of standard proteins were as indicated: 1. Rabbit muscle phosphorylase subunit (94,000), 2. Human plasminogen (87,000), 3. Bovine serum albumin (68,000), 4. Bovine ɤ-globulin heavy chain (50,000). 5. Light chain (23,500).

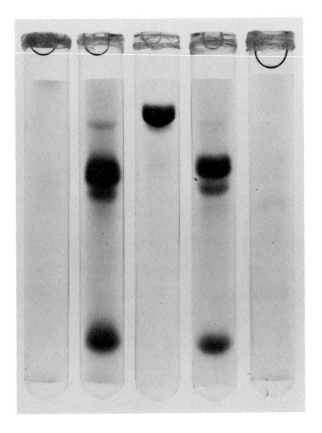

Figure 6   SDS Gel Electrophoresis of Human Plasminogen Activated by R2426 Plasminogen Activator and Urokinase:

Human plasminogen (500 µg) was incubated with either R2426 plasminogen activator or urokinase (200 units) in 25% glycerol and trasylol (2000 units) for 18 hr at 25°C (13). They were heated in the presence of mercaptoethanol (5%) and SDS (0.1%). From left to right: 1. R2426 plasminogen activator. 2. R2426 plasminogen activator + human plasminogen. 3. Human plasminogen. 4. Urokinase + human plasminogen. 5. Urokinase.

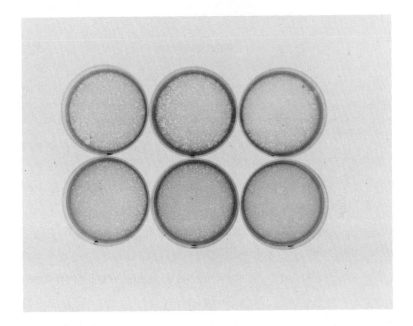

Figure 7 The effect of increasing concentration of purified anti-R2426 fibrinolysin on the "lytic" invasion by R2426 cells grown on fibrin plates. Antibody was incorporated in the fibrin prior to cell plating. Amount of fibrin per plate 1.5 ml, number of cells plated 2000. From left to right: Top (control): purified normal globulin 0.1, 0.2 and 0.3 ml. Bottom: purified antibody 0.1, 0.2 and 0.3 ml.

Discussion

A. Brown, National Cancer Institute: I may have missed this but have you assayed the tumor-bearing rats for activity of any of these products in their serum?

A. Yunis, University of Miami: The tumor itself?

A. Brown: Yes

A. Yunis: Yes, it has some activity.

A. Brown: In the whole serum of the rat can you separate a tumor specific protease?

A. Yunis: No, we have not studied serum from the rat.

J. Buchanan, Massachusetts Institute of Technology: Does your plasminogen-independent activity have any effect or even catalyze a slow conversion of plasminogen to plasmin. In other words does it have plasminogen activator activity?

A. Yunis: The problem here, Dr. Buchanan, is that we do not have the direct fibrinolysin completely free of plasminogen activator activity and therefore we cannot really say. That is what we are working on at the present.

J. Buchanan: For some time we have wondered whether the plasminogen-independent activity might have a dual role of activating plasminogen on the one hand and directly hydrolyzing fibrin clots on the other.

A. Yunis: That is a possibility, yes.

S. Magnusson, University of Århus: We have heard about several of the proteolytic enzyme inhibitors in blood/plasma at this meeting. There is one fairly recently discovered which has been described by Collen and collaborators from Louvain, Belgium which is an antiplasmin and is not identical with α-antitrypsin, α 2M or antithrombine-III and which reacts very quickly with plasmin - it seems to be an almost instantaneous reaction. I was wondering whether your tumor cell plasmin reacts with this anti-plasmin?

A. Yunis: No, I am sorry, I do not know.

ISOZYME COMPOSITION, GENE REGULATION AND METABOLISM OF EXPERIMENTAL HEPATOMAS*

S. WEINHOUSE, J.B. SHATTON AND H.P. MORRIS
The Fels Research Institute and Department of Biochemistry
Temple University School of Medicine;
Department of Biochemistry
Howard University

Abstract: There is an ever growing literature pointing to a misprogramming of genetic expression in cancer, based (a) on the ectopic production of polypeptide hormones by many human tumors of non-endocrine organs; and (b) on the disappearance in tumors of antigens expressed by the normal adult tissue of origin, and the appearance of tumor-associated neoantigens. A new functional dimension has been added to this concept by the recognition (c) that in cancer there is also a massive alteration in the isoenzyme composition. These findings are based largely on the use of the Morris hepatomas, a series of chemically-induced transplantable neoplasms which cover a wide range of differentiation, accompanied by corresponding differences in growth rate, and in their metabolic and enzymatic deviation from the adult liver. Tumors that are well differentiated and grow slowly retain to varying degrees those isoenzymes which are under host dietary and hormonal controls, and are kinetically programmed for specific hepatic function. However, with decreased differentiation and rapid growth, those liver marker isoenzymes are virtually lost; and are replaced by high activities of isoenzymes which are not under host control, have different regulatory properties, and are normally low or undetectable in the adult liver.

A striking feature is the appearance in the poorly differentiated hepatomas of isozymes that are low or absent in the differentiated tissue of origin but are the predominant or sole forms in fetal liver, or indeed in the whole early embryo. These changes exactly parallel alterations in antigen expression, and indicate that in cancer, genes that are active in fetal stages but are inactivated during normal embryonic development, are reactivated.

It appears that regulatory mechanisms which function in normal differentiation, and which rigidly determine both the site and temporal specificity of gene activation are impaired in cancer.

INTRODUCTION

In the more than half a century that has elapsed since Otto Warburg opened up the era of modern biochemical research in cancer, the cancer cell has been an insurmountable challenge. Theories on its origin and nature have been proposed, controversies have flared, hopes have risen and fallen, and as yet no chemical entity has been discovered whose presence or absence critically defines the cancer cell. In 1924, Warburg proclaimed that the answer to cancer was a defective respiration (1). Today, 50 years later, despite an unbelievable burgeoning of knowledge about both normal and neoplastic cells, the nature of cancer still eludes us. The broadening of our knowledge of cell biology, if it has taught us anything, it is that the problem of cancer is one of exceeding complexity.

Interwoven throughout the cancer literature is a common thread, however; that neoplasia is a disease of genetic regulation whose phenotypic expression is a mis-programming of protein synthesis. Evidence for this view comes from three sources.

TABLE 1

Molecular manifestations of abnormal gene expression in cancer

1. Ectopic polypeptide hormone synthesis.
2. Abnormal antigen expression.
3. Alterations of isozyme composition.

The clinical oncology literature contains numerous examples of the ectopic synthesis of various polypeptide hormones by non-endocrine tumors. Perhaps the most striking is the presence of adrenocorticotrophic hormone (ACTH) in the oat-cell tumor of the lung; but there are reports on the ectopic production of insulin, calcitonin, vasopressin, glucagon, parathormone, growth hormone, and many others by tumors of various types (2). Secondly, it is well established from the immunology literature that there is a loss in cancer of antigens present in normal adult differentiated tissue, while neoantigens are acquired that are absent from the adult tissue of origin, but may have been present in fetal organs (3). Thirdly, it is now becoming increasingly evident that similar aberrations occur in the isozyme composition of tumors.

A whole new dimension has been added to our understanding of enzyme regulation by the discovery by Markert in 1959 that enzymes can exist in multi-molecular forms (4). Termed "isoenzymes" or "isozymes" by Markert, the intervening years have seen such a tremendous burgeoning of knowledge in this field that the proceedings of the Third International Symposium on Isozymes, held last year in New Haven, required no less than four large volumes (5). These studies have had a powerful impact on biological science and medicine. In this presentation, I will describe and discuss changes in isozyme composition that are associated with the neoplastic process in liver.

The Morris hepatomas have been a most useful experimental model in studies of recent years on the metabolism and enzymology of tumors. These chemically induced, transplantable tumors represent a wide spectrum of growth rate and degree of differentiation.

TABLE 2

Characteristics of Morris hepatomas

Well differentiated
Slow growth
Normal or near normal karyotype
Retain hepatic enzymes and functions
Low aerobic glycolysis

Poorly differentiated
Rapid growth
Abnormal karyotype
Hepatic isozymes replaced by non-hepatic types
High aerobic glycolysis

Studies with these hepatomas have clearly demonstrated that tumors arising from a single cell-type of origin, namely the parenchymal liver cell, may exhibit great diversity in morphology, paralleled by diversity in growth rate and cellular metabolism (6,7,8). Many of the Morris hepatomas are poorly differentiated, grow rapidly, and are, therefore, very similar to the usual transplantable hepatomas which exhibit all of the functional activities of the cell of origin, and have a high rate of aerobic glycolysis. On the other hand, those Morris hepatomas, which by histologic criteria, are characterized as well or highly differentiated grow slowly, have chromosome number and karyotypes close to that of normal liver, and retain to varying degree some of the enzymatic and metabolic characteristics of the adult liver. A most striking feature of these slow-growing, well differentiated hepatomas is their low glycolytic activity (9,10). High glycolytic activity has always been considered to be a hallmark of the cancer cell; and this anomalous behavior by well-differentiated tumors provoked our interest. We recognized that these tumors, exhibiting as they do a wide range of glycolysis, not only provided a novel experimental system for exploring possible enzymatic bases for glycolytic regulation, but also for learning how the biological and metabolic aberrations of cancer are correlated with the loss or retention of those enzymes which play specific functional roles in normal liver metabolism. In this paper I would like to focus on one aspect of our work with these tumors over the past 15 years; namely the alterations of their isozyme composition; and to speculate on what these changes might be telling us about the nature of the neoplastic transformation.

Glucose-ATP Phosphotransferases. Figure 1 illustrates a typical pattern of isozyme alteration in this series of experimental tumors. Normal adult liver has four hexokinase isozymes. Three of these, collectively called hexokinase, and indicated here in the open bars, are in low activity, and are the same isozymes found widely distributed in various animal tissues besides liver. The major isozyme, termed glucokinase, and shown in the shaded bars, is virtually unique to liver and has kinetic properties which make it functionally important in hepatic glucose metabolism. Its activity is under rigid host hormonal control, being highly dependent on insulin. In diabetic animals, its activity is extremely low, but is rapidly restored by insulin administration. In highly differentiated hepatomas the isozyme pattern is identical with that of normal adult liver, with high glucokinase and low hexokinase activity. In well-differentiated, somewhat faster-growing hepatomas, glucokinase is sharply decreased, while hexokinases remain low. However, in the rapidly growing, poorly differentiated tumors, not only has glucokinase disappeared, but it has been replaced by a very high activity of the hexokinases (11). It is notable that the isozyme composition of the poorly differentiated hepatomas resembles that of fetal liver, which also has virtually no glucokinase. This isozyme does not appear in normal liver until two weeks after birth.

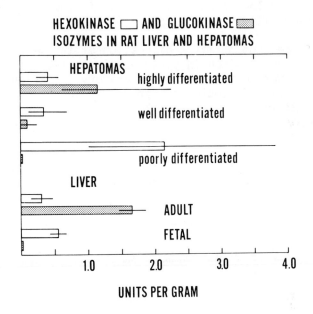

Fig. 1. Relative activities of glucose-ATP phosphotransferase isozymes in rat liver and hepatomas. Assay conditions in this and other figures were as described in the appropriate references.

Thus we see that with decreased differentiation and increased growth rate, there is a switch in isozyme composition, from that of normal adult liver in which an isozyme crucial to liver function predominates, to a pattern closely resembling that of fetal liver.

<u>Pyruvate Kinases</u>. An exactly parallel pattern was found with another key enzyme of glucose metabolism, namely pyruvate kinase (Figure 2). This enzyme operates at strategic crossroads of liver function, being a key step in glucose utilization and synthesis. In normal adult liver there is a preponderance of Type I, an isozyme which, like glucokinase, is under hormonal control by the host and is uniquely geared

for liver function. Again, we found that the highly differentiated hepatomas have the normal liver pattern. However, with decreased differentiation and increased growth rate the liver type is sharply decreased and is essentially completely replaced by an extremely high activity of Isozyme Type III, which is very low in adult liver, but is the sole form in fetal liver (12).

Time does not permit a discussion of the metabolic implications of this switch in pyruvate kinase isozyme pattern. However, experiments from our laboratory in various model systems (13,14,15) indicate that the presence of the high activity of Type III pyruvate kinase in the poorly differentiated hepatomas may be the key to their high glycolysis.

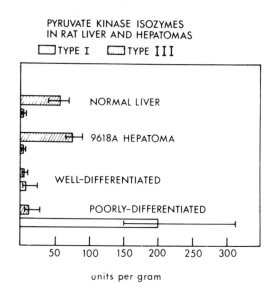

Fig. 2. Relative activities of pyruvate kinase isozymes in rat liver and hepatomas.

<u>Phosphorylases</u>. One more example taken from work of Dr. Sato in our laboratory (16,17) will suffice to illustrate the generality of this phenomenon. Glycogen phosphorylase exists in three immunologically distinct isozymic forms. The liver, as shown in Figure 3, has a high activity of the

liver type isozyme, and is unaccompanied by other forms. The 14-day whole rat embryo in contrast has mainly the fetal isozyme, together with a low activity of the adult muscle form. In hyperplastic nodules, which arise early in the treatment of rats with carcinogens, and which represent an early precancerous stage, the liver form predominates, but some fetal form now appears. With hepatomas of decreased differentiation and increased growth rate, the adult liver form progressively decreases, and in the poorly differentiated, fast growing hepatomas the fetal form now predominates.

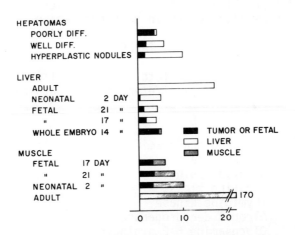

Fig. 3. Isozyme alterations in glycogen phosphorylase during embryonic development in the rat and in hepatomas.

The lower portion of Figure 3 shows how the phosphorylase isozyme composition changes during embryonic development. With liver development there is a steady increase of the adult liver type, and with muscle development there is a steady increase of the adult muscle isozyme. By two days after birth there is still some fetal isozyme present in both liver and muscle, but this ultimately disappears from both adult tissues.

Other Isozyme Alterations. A number of questions now emerge - first, how prevalent is this phenomenon? Our own experiments involve a limited series of enzymes of carbohydrate metabolism; however, there is a growing literature that points to a massive switch in isozyme composition as a general phenomenon in cancer. Listed in Table 3 are other examples. Fetal isozymes were found by us to replace the normal adult liver isozyme for aldolase, adenylate kinase, carbamyl phosphate synthetase, and more recently creatine kinase. Work of others has shown similar switches in the isozyme composition of the branched-chain amino acid aminotransferases, the glutaminases, thymidine kinases, alcohol dehydrogenases and glucosamine-6-phosphate synthase (7,8). It would appear that we have thus far observed the tip of an iceberg; a massive switch in gene activation.

TABLE 3

Enzymes in which fetal isozyme pattern occurs in poorly differentiated hepatomas

---

Aldolase
Branched chain amino acid aminotransferase
Glutaminase
Adenylate kinase
Creatine kinase
Thymidine kinase
Alcohol dehydrogenase
Glucosamine 6-P synthase

Illustrated in Figure 4 is a common pattern of divergence of isozyme composition accompanying the normal process of differentiation during normal embryonic development. A prototype isozyme, which is the sole or highly predominant form in the early embryo, is partially or completely lost as the adult organs develop. In cancer this process is reversed. It obviously does not occur necessarily with the neoplastic transformation since the well and highly differentiated hepatomas still retain the adult isozyme. The reversion to the fetal form is a step-wise process, which is not associated with all cancers; but is virtually complete in poorly differentiated tumors.

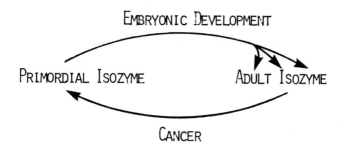

Fig. 4. Diagrammatic representation of reversal of normal differentiation in cancer.

Alterations of Antigen Expression. These isozyme alterations strikingly parallel a large body of immunology literature demonstrating analogous alterations of antigen expression in tumors. Tumors lose antigens characteristic of their tissue of origin and express neo-antigens that are not present in the adult normal tissues, but which may be present in fetal tissues (3). The presence of embryonal antigens has been documented in experimental hepatomas by Baldwin (18). Human liver tumors produce α-fetoglobulin, a protein present in fetal serum but not in postnatal serum (19); and a variety of gastrointestinal tumors produce a glycoprotein, the so-called carcinoembryonic antigen, which is absent from adult but is present in the fetal gastrointestinal tract (2). Their identification in human tumors is an exciting chapter of current cancer research, offering the prospect of early clinical detection.

Whether proteins are detected by their enzymatic or immunologic properties, there thus appears to be in cancer an inactivation of genes coding for proteins normally present in adult differentiated tissue; and a re-expression of genes active in the fetal stage but inactivated during embryonic development. The beautifully phased sequence of orderly development that occurs in normal differentiation is apparently reversed.

Ectopic Synthesis of Polypeptide Hormones. One of the most striking examples of misplaced protein synthesis in cancer is the bizarre production of polypeptide hormones by numerous non-endocrine neoplasms. Tumors of lung, prostate, kidney, stomach, liver and other organs produce a wide variety of such substances (21). These studies provide some basis for the belief that ectopic hormone production may also be embryonic in origin. Hall (2) has pointed to the common embryonic origin of the pituitary and lung in the endoderm, and suggests that the production of pituitary hormones by lung tumors is a reflection of this common embryonic ancestry.

Levine and Metz (22) have carried the analysis further and have classified most of the hormone-producing tumors in two groups.

TABLE 4

Tumors of Group I

Major Tumors
- Foregut carcinoid
- Oat cell carcinoma
- Islet cell tumors
- Pancreatic and biliary ducts
- Thyroid medullary carcinoma
- Malignant epithelial thymoma

Hormones

Insulin	Gastrin
Calcitonin	Glucagon
ACTH	Secretin
MSH	Vasopressin

TABLE 5

Tumors of Group II

Major Tumors

Hepatoma	Vascular
Cholangioma	Connective tissue and mesodermal
Wilms	Reticuloendothelial
Hypernephroma	Lung (except oat cell)
Adrenal cortical	Gastrointestinal
Non-germinal gonadal	Melanoma

Hormones

Parathormone	Growth hormone
Erythropoietin	Insulin-like
Gonadotropin	Renin
Placental lactogen	Thyrotropin
Prolactin	

Tumors of Group I arise from cells which originate embryologically in the neural crest. The tissues and the hormones that may be produced by their tumors are listed in Table 4. Tumors of Group II originate from the endoderm and mesoderm; the tissues and hormones produced by their tumors are listed in Table 5. Evidently cancers of a wide range of tissues can express potentialities possessed by cells from which they originated during normal embryonic development.

Another striking example of fetal isozyme expression is the ectopic presence in various human tumors and serum of an alkaline phosphatase normally present in the human placenta. Fishman (who discovered this so-called carcino-placental isozyme) and his coworkers have recently described a highly specific radioimmunoassay for its detection in human serum (23).

Possible Alterations of Other Proteins. In general, proteins can be identified immunologically, either as antigens or antibodies; or by their functional activity, as enzymes or hormones. By these criteria we find in cancer widespread alterations in protein composition. The significance of these to the neoplastic transformation still remains obscure; but there is little doubt that they contribute to neoplastic cell behavior. We have already noted that the replacement of key isozymes in hepatomas may be a causal factor in metabolic derangements, and may represent the molecular basis of their characteristically high aerobic glycolysis. The ectopic elaboration of hormones by human tumors of various types causes a multiplicity of systemic abnormalities that transcend those induced by the tumor itself. It is highly likely that similar abnormalities of protein synthesis occur; and the search for changes in the composition of nuclear proteins or in cell surface or intracellular membrane proteins may help in understanding some of the hitherto puzzling features of the cancer cell. Alterations in the protein composition of the cell surface for example not only might alter the antigen structure of the cell, rendering it unrecognizable to the immune systems, but might also lead to loss of reponsiveness to hormones and drugs, and to an impairment of the intricate mechanism whereby signals generated at the cell surface are transmitted to the machinery for protein synthesis and cell proliferation.

Molecular Basis of Abnormalities of Gene Expression. It is an article of faith in biology that all cells of an organism contain the same complement of genes. There is a relatively simple and elegant mechanism for selective gene expression operating in the bacterial cell. In such primitive systems there are linear arrays of regulatory and structural genes on the genome. Protein synthesis normally is repressed by a protein coded by a regulatory gene. There is an interplay between the repressor and co-repressors or inducers, which influences the binding of the repressor to the genome, and thereby determines whether a particular messenger RNA will be transcribed. Similar regulatory systems presumably operate in higher animals; and beautiful examples of hormonal and substrate control of enzyme synthesis are known in both normal and neoplastic cells (24).

A higher level of gene regulation must operate, however, in normal, adult differentiated tissue, where all but a minuscule portion of the genome is repressed. During normal embryonic development of the various organs, genetic expression is tightly and rigidly fixed and is essentially irreversible. It is this higher mechanism, which presumably determines the organ specificity of gene activity, and it is this mechanism that is apparently circumvented in cancer. This mechanism also may involve an exact juxtaposition of regulatory and structural genes, and, if so, it is not difficult to envision that this precise topographic arrangement might be altered by carcinogenic agents. With oncogenic viruses this could occur by integration of the viral genome at strategic sites; with radiation or chemicals, this could occur either by direct displacement or mutation of regulatory genes; or indirectly, by errors in their structure or placement introduced during repair of DNA damage. Granting such an injury, it would follow that regulatory mechanisms might be rendered unstable. Such an "unlocking" of the normally rigid specificity of gene repression could lead to a permissive condition that would allow inappropriate gene expression and thus lead to the various characteristics of neoplasia; the great diversity of phenotypes, misprogramming of protein synthesis, metabolic derangements, and uncontrolled cell proliferation. An unstable genome could also plausibly account for tumor progression, the sporadic and unpredictable succession of cellular alterations that inevitably culminate in anaplasia and unbridled growth.

Granting that in cancer we are dealing with a disorder of differentiation, several questions emerge. Differentiation involves an orderly programmed sequence. Is the reversion in cancer similarly ordered and programmed, as implied by such terms as derepressive, dedifferentiation, retrodifferentiation, etc. (3,7), or are the numerous expressions of fetal proteins a coincidental phenomenon - what Sugimura (25) has termed dis-differentiation?

It is important to recognize that the primordial isozymes are not strictly fetal, but may persist in many adult tissues, even though they are lost from skeletal muscle and liver. For example, the fetal forms of pyruvate kinase (12) and phosphorylase (16) are also present in adult kidney, brain, mammary gland and other tissues. Thus the isozymes that appear in poorly differentiated hepatomas are not necessarily just out of phase in respect to time, but rather are misplaced by site as well as by time. Endless questions need to be asked, but definitive answers must await new knowledge. It is my conviction that the phenomenon of misplaced protein synthesis in hepatomas is trying to tell us something important about the nature of the neoplastic transformation, but obviously we need to know more about the process of normal differentiation before we can understand this disorder of differentiation that may well lie at the heart of the cancer problem.

REFERENCES

(1) O. Warburg, The Metabolism of Tumours (Arnold Constable, London, 1930).

(2) L. Nathanson and T.C. Hall, Annals of The New York Academy of Sciences, 230 (1974) 367.

(3) N. Anderson and J.H. Coggin, Jr., in: Proceedings of the Second Conference on Embryonic and Fetal Antigens in Cancer (Oak Ridge National Laboratory, 1972) p. 361.

(4) C.L. Markert and F. Møller, Proc. Nat. Acad. Sci. U.S.A., 45 (1959) 753.

(5) C.L. Markert, editor, Isozymes (in 4 volumes) Academic Press, New York, 1975).

(6) H.P. Morris, in: Handbuch der Allgemeinen Pathologie (Springer-Verlag, Berlin, Heidelberg, New York, 1975) Vol. 1, p. 277.

(7) S. Weinhouse, Cancer Res., 32 (1972) 2007.

(8) T. Ono and S. Weinhouse, editors, Gann Monograph, Vol. 13: Isozymes and Enzyme Regulation in Cancer (University of Tokyo Press, Tokyo, 1972).

(9) A.C. Aisenberg and H.P. Morris, Nature 191 (1961) 1314.

(10) G. Weber, G. Banerjee and H.P. Morris, Cancer Res., 21 (1961) 933.

(11) J.B. Shatton, H.P. Morris and S. Weinhouse, Cancer Res., 29 (1969) 1161.

(12) F.A. Farina, J.B. Shatton, H.P. Morris and S. Weinhouse, Cancer Res., 34 (1974) 1439.

(13) C.H. Lo, V.J. Cristofalo, H.P. Morris and S. Weinhouse, Cancer Res., 28 (1968) 1.

(14) M. Gosalvez, J. Perez-Garcia and S. Weinhouse, Eur. J. Biochem., 46 (1974) 133.

(15) M. Gosalvez, L. Lopez-Alarcon, S. Garcia-Suarez, A. Montalvo and S. Weinhouse, Eur. J. Biochem., 55 (1975) 315.

(16) K. Sato, H.P. Morris and S. Weinhouse, Cancer Res., 33 (1973) 724.

(17) K. Sato and S. Weinhouse, Arch. Biochem. Biophys., 159 (1973) 151.

(18) R.W. Baldwin, Advances in Cancer Research, 18 (1973) 1.

(19) G.I. Abelev, Advances in Cancer Research, 14 (1971) 195.

(20) P. Gold, Prog. Exptl. Tumor Res., 14 (1971) 43.

(21) T.C. Hall, editor, Paraneoplastic Syndromes, Annals of The New York Academy of Sciences, 230 (1974).

(22) R.J. Levine and S.A. Metz, Annals of The New York Academy of Sciences, 230 (1974) 533.

(23) C.-H. Chang, S. Raam, D. Angellis, G. Doellgast and W.H. Fishman, Cancer Res., 35 (1975) 1706.

(24) G.D. Tomkins and T.D. Gelehrter, in: Biochemical Actions of Hormones (Academic Press, New York and London, 1972) p. 1.

(25) T. Matsushima, S. Kawake, M. Shibuya and T. Sugimura, Biochem. Biophys. Res. Comm., 30 (1968) 565.

*Work by the authors has been supported by Grants CA-12227, CA-10727, and CA-10916 from the National Cancer Institute and Grant BC-74 from the American Cancer Society.

Figures 1 to 3 are reproduced by courtesy of Cancer Research.

Discussion

B. Chance, University of Pennsylvania: Thank you very much Dr. Weinhouse. Do you get similar results in regenerating liver?

S. Weinhouse, Temple University: In some instances regenerating liver also shows these changes but not in most instances. For example with pyruvate kinase isozymes and glycogen phosphorylase isozymes, the normal pattern is retained even during the period of rapid regeneration of liver.

N. Kaplan, University of California: I wanted to ask if it is unusual for the creatine kinase to be present in the liver. Does it occur in any particular isozyme form in the undifferentiated hepatomas?

S. Weinhouse: Surprisingly, creatine kinase is present in liver and is retained in liver tumors. We have just begun this work and I am not really prepared to discuss it in any categorical way, but we have found creatine kinase not only in liver, but also in some non-hepatic tumors; also in muscle and mammary tumors. So far it appears that the tumor form is the form which is the embryo form.

N. Kaplan: Do you find creatine kinase in normal liver?

S. Weinhouse: Yes. As a matter of fact its activity is quite high although low compared with what you find in muscle. It is also widely prevalent in many tissues including brain, adipose, etc.

N. Kaplan: I have to bring up old skeletons. But I would like to ask what you think about the Warburg manifesto in the light of your present findings. Do tumor cells constantly continue to dedifferentiate, so that after carrying them for a long time, they are completely dedifferentiated with a great tendency toward aerobic glycolysis? Your data seem to fit the Warburg view. What do you think?

S. Weinhouse: I would regard the high aerobic glycolysis as not occurring with the initiation of tumors but rather a manifestation of a late stage of dedifferentiation. It is still a mystery why some primary tumors are highly differentiated and others are poorly differentiated. It is true however, that as tumors are successively transplanted, a highly differentiated, slow growing tumor which has a low aerobic glycolysis will gradually become less differentiated, will grow more rapidly and will take on a high aerobic-glycolysis. Sometimes this occurs more rapidly and will take on a high aerobic-glycolysis. Sometimes this occurs rather suddenly in a single generation, as we have seen this on several occasions with tumors that we have received from Dr. Morris, and this is irreversible.

R. Parks, Brown University: I wonder whether you have considered the possibility of applying some of the very striking differences that you have demonstrated between the fetal enzymes of tumor tissues and the adult enzymes of normal tissues for exploitation for the purpose of chemotherapy. We have an excellent example of the potential of this approach in the classic work of Burchall and Hitchings, who examined the enzyme dihydrofolate reductase from a wide variety of species ranging from bacteria, protozoa, helminths and other species up to mammals. These enzymes from this wide range of species resembled each other in most characteristics such as molecular weight, substrate specificity, cofactor requirements, etc. and all were inhibited to about the same extent by the folate antagonist methotrexate with K_i values in the order of 1×10^{-9}M. This is a tight-binding inhibitor that is very similar in structure to the natural substrates. However, the very important observation has been made that other inhibitors of the so-called "small molecule" class (mostly substituted diaminopyrimidines) varied greatly in their ability to inhibit dihydrofolate reductases from the various species. Perhaps the most interesting agent of this class is the drug, trimethoprim, which has a K_i value of about 1×10^{-9}M with various bacterial enzymes but a much less potent K_i value of about 1×10^{-4} to 10^{-5}M with mammalian dihydrofolate reductases. This dramatic difference in the inhibitory potency against the bacterial versus the mammalian enzymes is currently being exploited for clinical antibacterial chemotherapy especially since striking synergistic effects are seen when trimethoprim is employed in combination with drugs of the sulfanilamide class. I regard these findings of Burchall and Hitchings as among the most important conceptual advances in the field of chemotherapy of the past quarter century. It appears that structural differences in the enzyme dihydrofolate reductase not directly associated with the catalytic site of the enzyme (differences in enzyme structure probably resulting from many millions of years of evolution) have resulted in the binding sites on the bacterial enzymes that differ substantially from those of the mammalian enzymes. The observations reported by Dr. Weinhouse today may possibly present us with similar chemotherapeutic opportunities, since it is not unreasonable to expect that significant structural differences will be found between the fetal and adult enzymes. If such is the case, it should be possible to develop inhibitors that will specifically bind to the fetal enzymes of tumors more tightly than to the related adult enzyme. Therefore, I believe that these observations of Dr. Weinhouse present those of us who work in enzymology of cancer with both a challenge and an opportunity to seek out and hopefully find new and specific inhibitors of the fetal enzymes characteristic of tumor tissues. Has any progress been made along the lines that I have suggested?

S. Weinhouse: I agree that a knowledge of isozyme composition might well provide an additional dimension in chemotherapeutic research. It is obvious that mere enzyme assay can obscure profound differences in isozyme composition. Alerted to the possibility that tumors may contain isozymes which differ sharply from adult forms in kinetic properties, and in binding to substrates, cofactors, and inhibitors, etc., researchers in chemotherapy could not only approach the design of new drugs more intelligently but would also be better equipped to understand the action of drugs currently in use.

Other possible practical applications may be cited. If an antibody to a fetal isozyme could be prepared and attached covalently to an antitumor drug, it might conceivably seek out those tumor cells which contain the isozyme and thus provide a selectivity of effect on such tumors that have this isozyme. This approach would work probably well with isozymes associated with surface membranes.

If enzyme assays of sufficient sensitivity could be developed, it is possible that the detection of tumor-associated isozymes in the plasma might be helpful in early diagnosis and possible localization of tumors, in the same manner that coronary or brain lesions due to vascular obstruction are now diagnosed by plasma enzyme assays.

M. Horowitz, New York Medical College: Your work is derived from work with chemical carcinogenesis. Have you either in your laboratory or literature found similar generalization so far as isoenzyme pattern changes in virally induced tumors?

S. Weinhouse: No, we have not studied viral induced tumors. We have been sort of trapped by the liver because the liver does so many things. We have looked for differences between non-transformed and viral transformed cells; some sporadic efforts, but do not have good enzyme handles to explore isozyme changes in such cells.

B. Chance: May I interject another question? From what we know of the regulation of carbohydrate metabolism and the lack of strict relationships between the rates of glycolysis and the enzyme concentrations, do you think these changes of enzyme patterns are adequate to explain the increase of glycolytic activity?

S. Weinhouse: Yes, I think that the isozyme patterns can explain differences in glycolytic activity. I do not think that mere quantitative alterations in enzyme activities can do so because even with the most highly glycolizing tumor the enzyme activities are far in excess of the glycolytic activities. I think the key is regulation. Taking pyruvate kinase, as an example, the isozyme III which is the predominate or a sole form in highly glycolizing tumors has entirely different kinetic constants and allosteric effects, and these can certainly affect the rate of glycolysis.

B. Chance: Well it would be nice to document this point because I think it is pivotal to the hypothesis.

S. Weinhouse: In experiments with Dr. Mario Gonzalez we showed that by setting up model systems we could regulate the level of glycolysis and also the Pasteur effect by manipulating the relative activities of pyruvate kinase and mitochondrial respiration. We also showed that inhibitors of tumor pyruvate kinase will also affect the rate of glycolysis.

S. Grossman, Union Carbide Corporation: I am uncertain as to the use of the term "dedifferentiation", mechanistically. Do you suggest that an existing cell that is an adult functioning cell actually dedifferentiates, or that it is possible that a daughter cell fails to differentiate completely?

S. Weinhouse: I am not sure I heard everything you said, but I think you asked whether the highly differentiated cells become neoplastic, or whether undifferentiated cells become neoplastic and then redifferentiate.

S. Grossman: Yes, that is correct.

S. Weinhouse: Well, it is a good question, and I wish I knew the answer. I think that probably neoplasia can occur at different levels of cellular differentiation and I am not prepared to say at what level it is most likely to occur.

F. Huijing, University of Miami: The literature suggests that there are three types of phosphorylase: a muscle type, a smooth muscle type, and a liver type. Have you characterized your fetal enzyme far enough so that you could say whether it resembles the smooth muscle type, or could an alternative explanation be that it is the true liver phosphorylase because I have evidence in my laboratory that liver phosphorylase as it is isolated from adult liver is really partially digested proteolytically. The fetal enzyme gives much sharper bands on electrophoresis, it could be that it is not digested but the liver phosphorylase which gives broad bands is digested.

S. Weinhouse: Well, as far as we can tell, the so-called fetal or tumor phosphorylase is identical with the phosphorylase that occurs in kidney and in other normal adult tissues. It is immunologically entirely distinct from either the liver or muscle form.

F. Huijing: The smooth muscle form?

S. Weinhouse: Yes.

F. Huijing: But the liver form is not all that different from the smooth muscle form because like the smooth muscle phosphorylase it will form hybrids with the muscle enzyme.

S. Weinhouse: We have not studied smooth muscle per se so I am not sure of the identity there, but there appears to be identity among the tumor form, the fetal form, and the form that occurs in other non-hepatic or non-muscular tissues.

S. Grossman: I just thought I heard you say that glucokinase was under allosteric control? I was wondering what do you mean by that?

S. Weinhouse: Yes, I should not have said allosteric. It is under host control. It is well known that glucokinase is remarkably free of allosteric effects.

ENZYMATIC STRATEGY OF THE CANCER CELL

GEORGE WEBER

Laboratory for Experimental Oncology
Department of Pharmacology
Indiana University School of Medicine
Indianapolis, Indiana 46202

Abstract: This paper discusses our current approach to the knowledge of the enzymatic strategy of the cancer cell.

1. As a result of progress achieved through application of the conceptual and experimental approaches of the molecular correlation concept, the malignant transformation and progression were characterized by the behavior of key enzymes in a model system of liver tumors of different growth rates.

2. The progressive expression of neoplastic properties is revealed in the behavior of opposing key enzymes in carbohydrate, pyrimidine, DNA, purine and membrane cAMP metabolism.

3. There is an array of key enzymes that were shown to be markers of neoplastic transformation. These include the increased activities of glucose-6-phosphate dehydrogenase, transaldolase, glutamine PRPP amidotransferase, adenylosuccinate synthetase, adenylosuccinase and IMP dehydrogenase and UDP kinase. The activities that were decreased in all the tumors include those of the catabolic enzymes, xanthine oxidase, uricase and thymidine phosphorylase. It was recognized that the alterations in gene expression that resulted in the integrated imbalance of the various key enzyme activities confer selective advantages to the cancer cells.

4. The main genetic quantitative and qualitative alterations were tabulated as aspects of the ordered biochemical pattern that requires an explanation in terms of molecular mechanisms that would account for the meaningful nature of these alterations in cancer cells.

5. The molecular basis, biological significance and clinical relevance of the enzymatic strategy of the cancer cell were pointed out.

INTRODUCTION

The purpose of this paper is to examine the reprogramming of gene expression that is manifested in the enzymatic strategy of the cancer cell. This objective will be achieved by evaluating the recent advances achieved by application of the molecular correlation concept. The strategy of the cancer cell can be probed by exploring the pattern of gene expression as it is manifested in the pattern of concentration, activity and isozyme array of certain enzymes.

To gain insight into the strategy of malignancy it is useful to consider some of the main alterations that characterize neoplastic cells. These considerations (Table 1) draw attention to the biological, immunological, morphological and biochemical aspects of the emerging cancer cells.

TABLE 1

Malignancy-linked alterations

BIOLOGICAL ASPECTS:	*Continued replication*
	Increased growth fraction
	Invasiveness
	Metastasis
IMMUNOLOGICAL ASPECTS:	*Change in antigen expression*
	Escape from immunosurveillance
MORPHOLOGICAL ASPECTS:	*Altered cytology*
	Altered histology
BIOCHEMICAL ASPECTS:	*Transformation-linked imbalance*
	Progression-linked imbalance

Biological and clinical experience indicates that both transformation and progression play decisive roles in the clinical appearance of neoplasia. Therefore, the theoretical foundations of the molecular correlation concept required the fulfillment of two prerequisites in analyzing the biochemical aspects of neoplasia.

1. Biological Model System

A biological model system was needed which provided an array of transplantable neoplasms of the same cell types to permit a meaningful correlative analysis of the biological and biochemical behavior of the tumor cells. It was particularly relevant to the problems to be solved that the model system meets the following two criteria. (A) All lines should be malignant neoplasms, thus permitting examination of the transformation-linked events which appear in all the tumors as all-or-none characteristics of the neoplastic transformation. Such alterations should be indicators of transformation-linked events. (B) There was a need to have an array of neoplastic lines in which the different degrees in the expression of malignancy could be studied. Thus, the metabolic alterations that would be connected positively or negatively with the degrees of malignancy would be indicators of the process of malignant progression.

There are a number of tumor lines from which such a tumor spectrum can be assembled. The biological-biochemical framework of my investigations was initially carried out in a series of liver tumors of different growth rates that provided a spectrum of hepatocellular carcinomas of different malignancy (1-4).

2. Biochemical Discriminants

It was of critical importance that for the study enzymes be selected through which the pattern of gene strategy was expressed and regulated. I termed such enzymes "key enzymes" (2-5, 8). Studies in my Laboratory demonstrated that the strategy of gene expression in normal and neoplastic cells was manifested in the pattern of reciprocal behavior of opposing key enzymes in antagonistic and competing synthetic and degradative pathways (2-9). As a result of our investigations the conclusion was reached that identification and analysis of behavior of key enzymes and their relationship to the malignancy of the neoplasms permits evaluation of the linking of biochemical alterations with the biological behavior of the tumors.

(i) <u>Biochemical behavior stringently linked with neoplastic behavior.</u> Our investigations indicated that the activity, concentration and isozyme pattern of key enzymes and metabolic pathways may be linked with neoplastic (a) transformation and (b) progression.

(ii) Biochemical behavior not linked with neoplasia. The results of our studies also indicated that there are enzymes the behavior of which apparently is not linked with the core of neoplasia. Such enzymes usually are present in excess and govern reversible reactions.

Recognition of Ordered Alterations in the Biochemistry of the Cancer Cell

Our studies revealed that there was an ordered pattern of biochemical imbalance that could be detected in the model system of tumors of different growth rates by examining the key enzymes and the opposing pathways of synthesis and catabolism. We observed that the biochemical imbalance can best

TABLE 2

The ordered alterations in the biochemistry of the cancer cell can be detected by examination of behavior and relation to malignancy of:

---

a. Key enzymes that oppose each other in antagonistic pathways

b. Opposing pathways of synthesis and catabolism

c. Ratios of opposing key enzymes and pathways

d. Isozyme shift

---

be detected by evaluation of the ratios of opposing key enzymes and pathways. In addition to the quantitative alterations, the reprogramming of gene expression in cancer cells also exhibited qualitative alterations as manifested in change of isozyme pattern which I term "isozyme shift" (3) (Table 2).

Identification of Key Enzymes

A conceptual advance was achieved when it was recognized that not all enzymes should be expected to relate to transformation and progression but only the key enzymes through which the reprogramming of gene expression operates (4-9,11). For this reason in Table 3 the main criteria I have found useful in identifying the key enzymes are given (4,8).

TABLE 3

Identification of key enzymes

FEATURES OF KEY ENZYMES	EXAMPLES IN VARIOUS METABOLIC AREAS
I. PLACE IN PATHWAY	
1. First in a reaction sequence	Pyruvate carboxylase; thymidine kinase
2. Last in a reaction sequence	DNA polymerase; Glucose-6-phosphatase
3. Pathways in themselves	Glucose-6-phosphatase; adenylate cyclase
4. Operate on both sides of reversible reaction pools	Thymidine kinase; dihydrothymine dehydrogenase
5. One-way enzyme opposed by another one-way enzyme	Phosphofructokinase; fructose diphosphatase
II. REGULATORY PROPERTIES	
6. Possesses relatively low activity in the pathway	Pyruvate carboxylase; DNA polymerase
7. Rate-limiting in pathway	PEP carboxykinase; ribonucleotide reductase
8. Target of feedback regulation	Thymidine kinase; Glutamine PRPP amido-transferase
9. Target of multiple regulation	Pyruvate kinase; ribonucleotide reductase
10. Exhibits allosteric properties	Aspartate carbamyltransferase
11. An interconvertible enzyme	Glycogen phosphorylase
12. Has isozyme pattern	Glucokinase-hexokinase, pyruvate kinase
III. BIOLOGICAL ROLE	
13. Involved in overcoming thermodynamic barriers	Four key gluconeogenic enzymes
14. Final common path of two or more metabolic pathways	DNA polymerase; glucose-6-phosphatase

Biochemical Imbalance in Different Metabolic Areas

In systematic studies over the past two decades strategic areas of intermediary metabolism have been explored in testing the proposal that the reprogramming of gene expression in the cancer cell is manifested and can be best detected in the behavior of key enzymes. The main metabolic areas are summarized in Table 4. These studies led to the conclusion that an ordered pattern of biochemical imbalance is revealed in the progression-linked and transformation-linked alterations of key enzymes, isozymes and opposing and competing metabolic pathways (Table 4).

TABLE 4

The biochemical imbalance in the hepatoma spectrum was characterized in the following metabolic areas:

Carbohydrate (Gluconeogenesis, glycolysis)

Pentose phosphate (Oxidative and non-oxidative pathways)

Pyrimidine (De novo and salvage pathways)

Purine (IMP synthesis, degradation and utilization)

Urea cycle

Ornithine utilization

Polyamine biosynthesis

Membrane cAMP synthesis and degradation

Protein and amino acid

Conclusion: An ordered pattern is revealed in:

Progression-linked alterations* (Class 1)

Transformation-linked alterations* (Class 2)

---

*In activities of key enzymes, isozymes, and opposing metabolic pathways.

BIOCHEMICAL STRATEGY OF THE CANCER CELL: PLEIOTROPIC ALTERATIONS MANIFESTED IN THE REPROGRAMMING OF GENE EXPRESSION

As a result of the conceptual and experimental advances achieved through application of the molecular correlation concept, it became possible to group the indicators of gene expression according to their linking with malignant behavior (Table 5). Class 1: Activities that correlate positively or negatively with tumor malignancy indicate reprogramming of gene expression that is connected with the different degrees in the expression of the neoplastic transformation. Class 2: Activities that are altered in all hepatomas indicate reprogramming of gene expression that is linked with the malignant transformation per se. Class 3: In turn, biochemical activities that do not relate to transformation or progression are recognized as random, coincidental changes that are not linked with the core of neoplasia.

Progression-Linked Alterations

A great deal of our investigations concerned the molecular correlates of the degrees in the expression of malignancy, i.e., the progression-linked alterations. These studies revealed the operation of an imbalance in the activities of opposing key enzymes and metabolic pathways in carbohydrate, pyrimidine, DNA, ornithine and membrane cAMP metabolism (1-4, 6, 9-12). Some of the reciprocal alterations that are linked with the degrees in expression of neoplastic transformation are summarized in Table 6. This is a sample of the over 70 biochemical parameters that so far have been shown to correlate with malignancy and growth rate in the hepatoma spectrum (2-4, 9, 12).

Transformation-Linked Alterations

In this Laboratory recently there were discovered 8 transformation-linked alterations in key enzyme activities that occur in all hepatomas irrespective of growth rate, malignancy and the degree of histological differentiation. Critical evaluation of the pattern of these alterations revealed that this reprogramming of gene expression produced an increased capacity to channel precursors to strategic biosynthetic processes and a decreased ability to degrade or recycle such precursors.

TABLE 5

Biochemical strategy of the cancer cell: pleiotropic alterations manifested in the reprogramming of gene expression

LINKING OF BIOCHEMICAL ALTERATIONS WITH MALIGNANCY	INDICATORS OF REPROGRAMMING OF GENE EXPRESSION: ACTIVITIES OF ENZYMES, ISOZYMES, PATHWAYS
CLASS 1 PROGRESSION-LINKED DISCRIMINANTS*	Activities that *correlate* with tumor malignancy: indicate reprogramming of gene expression that is linked with the different degrees of expression of the neoplastic transformation
CLASS 2 TRANSFORMATION-LINKED DISCRIMINANTS*	Activities that are increased or decreased *in all hepatomas*: indicate reprogramming of gene expression that is linked with the malignant transformation per se
CLASS 3 COINCIDENTAL ALTERATIONS	Activities that do not relate to growth rate: random alterations not connected with neoplastic transformation or progression

*Biochemical parameters that discriminate the pattern of liver neoplasia from that of normal, fetal, differentiating or regenerating liver.

TABLE 6

Reciprocal alterations linked with degrees of expression of neoplastic transformation

FUNCTIONS INCREASED		FUNCTIONS DECREASED	
PATHWAYS	ENZYMES	PATHWAYS	ENZYMES
CARBOHYDRATE METABOLISM			
Glycolysis	Hexokinase Phosphofructokinase Pyruvate kinase	Gluconeogenesis	Glucose-6-phosphatase FDPase PEP CK Pyruvate carboxylase
DNA METABOLISM			
De novo pathway	Ribonucleotide reductase dCMP deaminase dTMP synthase		
Salvage pathway (TdR to DNA)	TdR kinase dTMP kinase DNA polymerase	Thymidine degradation (TdR to CO_2)	Dihydrothymine DH
UMP biosynthesis	Asp. transcarbamylase	UMP degradation	Dihydrouracil DH
ORNITHINE METABOLISM			
Ornithine to polyamine synthesis	Ornithine decarboxylase	Ornithine to urea cycle	Ornithine carbamyl transferase
cAMP METABOLISM IN MEMBRANE			
Depletion in cAMP	cAMP phosphodiesterase	cAMP synthesis	Adenylate cyclase

The opposing pathways of glycolysis and gluconeogenesis and the oxidative and non-oxidative pathways of pentose phosphate metabolism in hepatomas are indicated in Figure 1. Our investigations showed that the key enzymes of glycolysis, hexokinase, phosphofructokinase, and pyruvate kinase, increased whereas the opposing key gluconeogenic enzymes, glucose-6-phosphatase, fructose-1,6-diphosphatase, phosphoenolpyruvate carboxykinase and pyruvate carboxylase, decreased with the increase in tumor malignancy (1-4). In turn, it was observed that the activity of the key enzyme of the oxidative pathway, glucose-6-phosphate dehydrogenase, and that of the key enzyme of the non-oxidative pathway, transaldolase, were increased in all the liver tumors (4, 6, 13). The elevation in all hepatomas of the activities of the dehydrogenase and the transaldolase increases the potential for routing glycolytic intermediates into pentose phosphate biosynthesis. Subsequent work showed that the enzyme linking pentose phosphate metabolism with purine biosynthesis, PRPP synthetase, was increased in the rapidly growing hepatomas (14). This observation was in line with the heightened capacity to produce purines in liver neoplasia. Recent investigations demonstrated that the activity of the first enzyme committed to purine biosynthesis that utilizes PRPP for de novo purine biosynthesis, glutamine PRPP amidotransferase, was increased in all the liver tumors (15). The rise in amidotransferase activity should result in an increased potential for de novo IMP biosynthesis (15, 16). In turn, the rate-limiting enzyme of IMP catabolism, xanthine oxidase, (16, 17) and the final enzyme of purine degradation, uricase, (16) were decreased in activity in all the hepatic neoplasms. These alterations should yield a decreased degradative capacity in purine metabolism. As a result of the increase in the activity of the key purine-synthesizing (amidotransferase) and a decrease in the activity of the rate-limiting purine-catabolizing enzyme (xanthine oxidase), the ratio of the synthetic to the catabolic enzyme increased roughly in parallel with tumor growth rate (Table 7). These experimental results demonstrated an imbalance in the pattern of key enzymes of IMP metabolism and directed our attention to the metabolic fate of IMP and the behavior of the enzymes involved in IMP utilization.

The enzymes immediately involved in IMP utilization, IMP dehydrogenase that produces XMP and adenylosuccinate synthetase that leads to SAMP, were examined as possible indicators of the capacity for synthetic utilization of IMP. In addition, adenylosuccinase, that utilizes SAMP for biosynthesis of AMP, was also studied (16, 18-20). The outcome of these

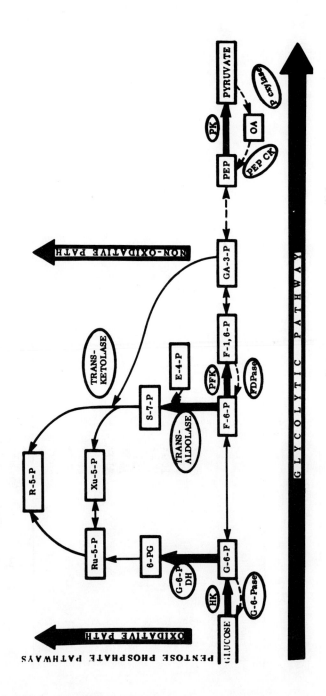

Fig. 1. Effect of neoplasia on activities of key enzymes of glycolysis, gluconeogenesis and pentose phosphate pathways. Increased activities are shown with heavy arrows; decreased activities with interrupted arrows.

TABLE 7

Antagonistic behavior of opposing key enzymes of IMP synthesis and degradation in hepatomas

Tissues	Amidotransferase activity	Xanthine oxidase activity	Amidotransferase/ xanthine oxidase ratios
Normal liver	100	100	100
Hepatomas			
9618A	218*	32*	689*
20	217*	50*	412*
9618B	169*	16*	1050*
16	146*	10*	1469*
47C	251*	22*	1333*
28A	174*	48*	357*
8999	319*	50*	666*
9633	153*	58*	271*
7800	253*	61*	450*
3924A	207*	10*	1937*
3683	280*	10*	2740*

Specific activities (μm/hr/mg protein × 10^{-2}) *of glutamine PRPP amidotransferase* (6.9 ± 0.1) *and xanthine oxidase* (10.8 ± 0.4) *yield ratios for AT/XO* = 0.6 ± 0.03.
*Statistically significantly different from values of normal rat liver

investigations showed that the activities of adenylosuccinate synthetase and adenylosuccinase were increased in all the hepatomas. In turn, the activity of IMP dehydrogenase was increased in all the liver neoplasms and the rise in activity correlated positively with tumor growth rate. The increased activities in the enzymes that utilize IMP for the biosynthesis of adenine and guanine nucleotides should provide a stepped-up ability of the cancer cells for the synthetic utilization of purines (16, 18-20).

KEY PURINE METABOLIZING ENZYMES: PROPERTIES AND RELATION TO NEOPLASIA

Comparison of activities and kinetic properties of the key purine metabolizing enzymes indicated an important relationship of the enzymes with neoplasia. Table 8 demonstrates that among all the enzymes of purine metabolism IMP dehydrogenase exhibits the lowest activity. Among the synthetic enzymes adenylosuccinase has the highest activity. From this Table emerges a relationship of the purine synthesizing enzymes that suggests that the synthetic enzymes which have the lowest activities in the normal resting liver are the ones that increase to the highest extent in the rapidly growing liver tumors. For instance, in the rapidly growing hepatoma 3683 the activity of IMP dehydrogenase was elevated 13-fold, but that of the adenylosuccinase only increased 1.75-fold (16). This relationship is similar to the one we recognized earlier for enzymes of pyrimidine and DNA biosynthesis (11). There also the enzyme activity that is the lowest (ribonucleotide reductase) increased to the greatest extent, whereas the activity of the enzyme with the highest activity in the resting liver (nucleoside diphosphate kinase) showed the smallest increase (11).

Recent studies demonstrated that the activity of UDP kinase was increased in all the liver tumors (21) and earlier work indicated that TdR phosphorylase was decreased in all the hepatomas. The enzymatic markers of malignant transformation are summarized in Table 9.

These elevations in all the hepatomas in activities of key enzymes of ribose-5-phosphate and purine biosynthesis, IMP utilization and UTP production, along with the decrease in the activities of catabolic enzymes, indicate a reprogramming of gene expression that is specific to malignant transformation (2-4, 22) and should confer selective biological advantages to the cancer cells.

TABLE 8

Comparison of activities of enzymes of the anabolic and catabolic pathways of purine metabolism

Enzymes	E.C. number	Substrate K_m (μM)	Livers Normal fed		Hepatomas Rapidly growing (3683)
			μm/hr/g wet weight	pm/hr/mg protein	% liver
ANABOLIC ENZYMES					
IMP dehydrogenase	1.2.1.14	12	0.19	2,140	1,292
GMP synthetase	6.3.4.1		1.70	19,000	
SAMP synthetase	6.3.3.4	36	3.10	36,000	370
Glutamine PRPP amidotransferase	2.4.2.14	900	6.00	60,000	278
Adenylosuccinase	4.3.2.2	5	32.00	365,000	175
CATABOLIC ENZYMES					
Xanthine oxidase	1.2.3.2	0.7	10.00	100,000	10
Uricase	1.7.3.3	100	150.00	645,000	5

From (16).

TABLE 9
Enzymatic markers of malignant transformation (4)

METABOLIC AREA	SYNTHETIC PATHWAY: INCREASED	DEGRADATIVE PATHWAY: DECREASED
PENTOSE PHOSPHATE	GLUCOSE-6-PHOSPHATE DEHYDROGENASE	
	TRANSALDOLASE	
PURINE	GLUTAMINE PRPP AMIDOTRANSFERASE	XANTHINE OXIDASE
	IMP DEHYDROGENASE	URICASE
	ADENYLOSUCCINATE SYNTHETASE	
	ADENYLOSUCCINASE	
DNA	UDP KINASE	TdR PHOSPHORYLASE

INTEGRATED PATTERN OF IMBALANCE IN CARBOHYDRATE, PENTOSE PHOSPHATE AND PURINE METABOLISM

Figure 2 provides a picture of the integrated imbalance in the activities of key enzymes of carbohydrate, pentose phosphate and purine metabolism in the hepatoma cells.

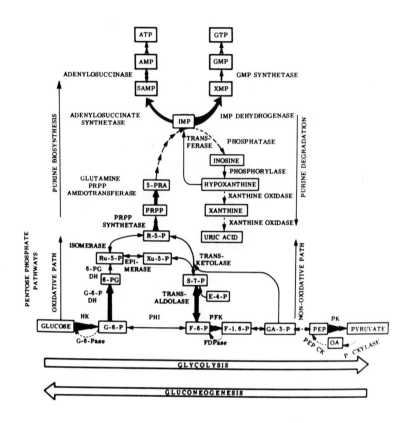

Fig. 2. Integrated imbalance of key enzymes of carbohydrate and purine metabolism.

MOLECULAR MECHANISMS THAT MAY UNDERLIE THE ORDERED BIOCHEMICAL PATTERN OF THE CANCER CELL

A consideration of the ordered and meaningful biochemical imbalance that was discovered in the spectrum of hepatomas of different growth rates leads to the inquiry regarding the molecular mechanisms through which the reprogramming of gene expression is accomplished by the cancer cell. In Table 10 are summarized some of the key events that require an account of mechanisms at the molecular level.

TABLE 10

Molecular mechanism must account for the following ordered biochemical pattern in the strategy of the cancer cell

GENETIC:	HEREDITABILITY
	GENE POOL STABILITY
	IRREVERSIBILITY OF IMBALANCE
	INTEGRATED MULTIENZYME IMBALANCE
QUANTITATIVE:	ANTAGONISTIC CHANGES
	GRADED QUANTUM JUMPS
	PROGRESSION-LINKED
	TRANSFORMATION-LINKED
QUALITATIVE:	QUALITATIVE ALTERATIONS: ISOZYME SHIFT
	ALTERED RESPONSES OF ENZYME ACTIVITY TO CONTROLS
	ALTERED RESPONSES OF ENZYME AMOUNT TO REGULATION
	SPECIFICITY TO NEOPLASIA

The mechanism proposed should account for the fact that the biochemical pattern is hereditable, and that the gene pool is stable as manifested in the constancy of the biological malignancy and the biochemical pattern of the different tumor lines. The mechanism should account for the phenomenon that a biochemical imbalance is irreversible and it is revealed in an integrated, multi-enzyme imbalance. The poly-enzymic nature of the biochemical pattern is particularly relevant, since a number of these enzymes are coded in genes located in different chromosomes. There is a requirement for a molecular mechanism that can account for biochemical changes that are antagonistic, that exhibit graded alterations and might be progression- or transformation-linked. It is of particular interest that there are qualitative alterations in this pattern, as expressed in the isozyme shift and in the altered responses to control of the activity and the amount of the various enzymes. Finally, the mechanism should account for the observation that the biochemical pattern of the hepatomas is specific to the neoplastic transformation and no similar pattern has been observed in fetal, differentiating or normal liver.

The main possibilities for the alterations have been discussed in terms of genetic or epigenetic mechanisms and these considerations were discussed in detail elsewhere (6). I point out that with the availability of a meaningful model system and the identification of the key enzymes and the linking of their behavior with neoplastic transformation and progression, the road should now be open for a critical enquiry into the molecular mechanisms that led to the display of the ordered pattern of biochemical imbalance in neoplastic cells.

ENZYME ACTIVITY AND ENZYME PATTERN AS INDICATORS OF REPROGRAMMING OF GENE EXPRESSION: CLINICAL RELEVANCE

With the progress in our understanding of the role of key enzymes and their linking to transformation and progression, it is relevant to underline the biological significance and clinical relevance of our observations (Table 11).

The progression-linked alterations in activities of key enzymes, in isozyme pattern and in the behavior of metabolic pathways can be used as indicators of the degrees of malignancy and they should be of assistance in the biochemical grading of neoplasms.

The transformation-linked alterations are markers of malignancy and they should be useful in the biochemical diagnosis of neoplasia.

TABLE 11

Enzyme activity and isozyme pattern: indicators of reprogramming of gene expression

MOLECULAR BASIS	BIOLOGICAL SIGNIFICANCE	CLINICAL RELEVANCE
1. PROGRESSION-LINKED ALTERATIONS	INDICATORS OF DEGREES OF MALIGNANCY	GRADING
2. TRANSFORMATION-LINKED ALTERATIONS	MARKERS OF MALIGNANCY	DIAGNOSIS
3. KEY ENZYMES IN TUMOR	TARGETS OF DRUG ACTION	SELECTIVE CHEMOTHERAPY
4. KEY ENZYMES IN HOST TISSUES	TARGETS OF SIDE EFFECTS	TOXICITY
5. DRUG-INDUCED CHANGE IN ENZYMES	CHEMICAL EVALUATION OF DRUG ACTION	THERAPEUTIC EFFECTIVENESS

The key enzymes in the tumor tissue should provide sensitive targets to anti-cancer drugs and thus the pinpointing of these enzymes and their behavior in tumors should be relevant in the design of selective chemotherapy.

The key enzymes in the host tissues are targets of side effects and their study should provide useful information regarding the extent and mechanism of toxicity.

The drug-induced changes in enzyme concentration should be helpful in the biochemical evaluation of drug action and should provide a molecular measure of the therapeutic effectiveness of the various anticancer compounds.

ACKNOWLEDGMENT

The research work outlined in this paper was supported by USPH NCI grants CA-05034 and CA-13526.

REFERENCES

1. G. Weber, Advan. Cancer Res., 6 (1961) 403.

2. G. Weber, Advan. Enzyme Regulat., 11 (1973) 79.

3. G. Weber, in: The Molecular Biology of Cancer, ed. H. Busch (Academic Press, New York, 1974) p. 487.

4. G. Weber, Biochemical Strategy of the Cancer Cell (1976) to be published.

5. G. Weber, R. L. Singhal and S. K. Srivastava, Advan. Enzyme Regulat., 3 (1965) 43.

6. G. Weber, A. Trevisani and P. C. Heinrich, Advan. Enzyme Regulat., 12 (1974) 11.

7. G. Weber., Israel Journal Med. Sci., 8 (1972) 325.

8. G. Weber, in: Mechanism of Action and Regulation of Enzymes, Proc. Ninth FEBS Meeting (1975) 237.

9. J. A. Ferdinandus, H. P. Morris and G. Weber, Cancer Res., 31 (1971) 550.

10. H. G. Williams-Ashman, G. L. Coppoc and G. Weber, Cancer Res., 32 (1972) 1924.

11. G. Weber, S. F. Queener and J. A. Ferdinandus, Advan. Enzyme Regulat., 9 (1971) 63.

12. G. Weber, S. F. Queener and H. P. Morris, Cancer Res., 32 (1972) 1933.

13. L. E. Selmeci and G. Weber, FEBS Letters (1976) in press.

14. P. C. Heinrich, H. P. Morris and G. Weber, FEBS Letters, 42 (1974) 145.

15. N. Prajda, N. Katunuma, H. P. Morris and G. Weber, Cancer Res., 35 (1975) 3061.

16. G. Weber, N. Prajda and R. C. Jackson, Advan. Enzyme Regulat., 14 (1976) in press.

17. N. Prajda and G. Weber, FEBS Letters, 59 (1975) 245.

18. R. C. Jackson, H. P. Morris and G. Weber, Biochem. Biophys. Res. Comm., 66 (1975) 526.

19. R. C. Jackson, H. P. Morris and G. Weber, (1976) submitted for publication.

20. R. C. Jackson, G. Weber and H. P. Morris, Nature, 256 (1975) 331.

21. J. C. Williams, H. P. Morris and G. Weber, Nature, 283 (1975) 567.

22. G. Weber, in: Liver Regeneration after Experimental Injury, eds. R. Lesch and W. Reutter (Stratton Intercontinental Medical Book Corp., New York, 1975) p. 103.

Discussion

L. Menahan, Medical College of Wisconsin: You eluded to changes in enzyme levels but you have not referred in any of your slides to metabolite levels. And if indeed the flux through the hexose monophosphate shunt is increased, how is the NADPH consumed that is produced through this cycle? Is there an increase in transhydrogenase or fatty acid synthesis in the hepatomas?

G. Weber, Indiana University: I refer to increased enzyme activities because this statement is supported for a number of enzymes by independent lines of evidence, by determination of the enzyme activity under linear conditions and measurement of enzyme concentration by immunotitration (G. A. Dunaway, Jr., H.P. Morris and G. Weber, Cancer Res. 34(1974) 2209; L.E. Selmeci and G. Weber, FEBS Letters (1976) in press). For the various metabolic parameters and metabolites we measured concentrations and flux by freeze clamp studies and by isotope techniques. Some of this work was done at Indiana Unviersity (G. Weber, M. Stubbs and H.P. Morris, Cancer Res. 31 (1971) 2177), some of it in Oxford, England, in collaboration with Sir Hans A. Krebs and his associates (D.H. Williamson, H.A. Krebs, M. Stubbs, M.A. Page, H.P. Morris and G. Weber, Cancer Res. 30 (1974) 2049. Our studies demonstrated that there is an increased glycolytic flux in the hepatomas that correlated with tumor growth rate and we also brought evidence for an increased activity of the direct oxidative pathway which correlated positively with tumor growth rate (M.J. Sweeney, J. Ashmore, H.P. Morris and G. Weber, Cancer Res. 23 (1963) 995). The evidence indicates that the increased production of NADH is not linked with increased fatty acid production, since the conversion of glucose to fatty acids was markedly decreased in all liver tumors we examined (Ibid). It is likely that the NADPH produced is utilized in DNA and RNA biosynthesis and in the folate reductase pathway.

L. Menahan: I just wanted to know if indeed this folate reduction pathway is significant enough to account for the increased flux of the NADPH produced by the hexose monophosphate shunt.

G. Weber: We do not have precise measurements of the NADP and NADPH concentrations and we do not have a complete accounting for the utilization of NADPH as yet.

S. Grossman, Union Carbide Corporation: I am wondering - why do you suppose we don't get an increase in the hypoxanthine guanine phosphoribosyl transferase, the purine salvage enzyme in hepatoma?

G. Weber: Current studies indicate that there is no change in the activity of the salvage enzyme in the hepatomas. One reason that it does not have to change in concentration is because it has the highest activity among the purine synthetic enzymes; it is present in excess. Moreover, this enzyme has very high affinity for the substrate so that it can readily recycle it to IMP. I pointed out elsewhere that such enzymes, that are present in an excess, are not linked either with transformation or with tumor growth rate or progression and they are grouped in Class 3 by the molecular correlation concept (G. Weber, in The Molecular Biology of Cancer (H. Busch, ed.) (1974) 487).

H. Morris, Howard University: I would like to comment on the stability of the tumor lines that Dr. Weinhouse and Dr. Weber have alluded to in their talks. I believe the initial transformation of the cancer cell in these different lines has remained reasonably stable. We have transplanted some of these poorly differentiated tumors for 25 years and they still retain their approximate growth rate. We have transplanted other highly differentiated lines for maybe 15 years and they still retain the same approximate growth rate. Now there have been some changes as Dr. Weinhouse mentioned in which there has been progression to very rapidly growing poorly differentiated tumors - but we feel that we have been able to maintain the respective tumor lines in a very stable situation. Now there may be some faster growing cells that are mixed with the slow growing tumors and it may be that these became more rapidly growing tumor lines. Once they do that, they never go back to slow growing lines and I feel that this is an evidence of progression of malignancy and is something that we have to try to delineate as Dr. Weber has presented this morning.

G. Weber: I think this important comment of Dr. Morris underlines the fact I referred to in that we are dealing here in the various tumor lines with fairly stable gene pools. The biochemical and histological studies we have been doing in all the hepatoma lines confirm that these tumor lines are fairly stable. As I said, this interesting phenomenon requires an explanation at the molecular level. This stability is one of the important properties that makes these tumor lines such particularly useful biological model systems.

G. Koch, Roche Institute of Molecular Biology: I refer to a recent paper by Ames in the Nov. issue of PNAS (T.C. Stephens et al. PNAS 72, 4389-4373, 1975). Small effector molecules like pp G pp (guanosine 5' diphosphate, 3' diphosphate) affect profound regulatory functions on both anabolic and catabolic pathways. Are you willing to speculate or ready to add to your extensive and impressive list on possible regulatory mechanisms at the level of transcription and translation the possibility that coordinated increases and decreases in enzyme levels for anabolic and catabolic pathways in your system might be triggered also by small effector molecules?

G. Weber: I have attempted to do this in reviewing the possible levels of regulation in one of my slides and perhaps this is not the time to go any further into this problem as yet. Did you refer to a microbial system in your comments?

G. Koch: Yes. The observations were obtained in studies employing S. typhimurium.

G. Weber: Yes, you see that with the availability of the model systems of hepatomas of different growth rates I do not feel that microbial systems provide any advantage or particularly meaningful pattern of comparison.

R. Parks, Brown University: Several years ago Elford and colleagues (J. Biol. Chem. 245, 5228, 1970) on the basis of studies with a variety of hepatomas isolated by Dr. Morris that varied greatly in their rates of growth, reported a striking correlation between the tumor growth rate and the activity of the enzyme ribonucleotide diphosphate reductase. On the basis of these findings, Elford postulated that this enzyme is the rate-limiting one in cellular growth. This observation has been widely discussed and has been accepted to the point that it is now found in some of the more prominent textbooks. How do Elford's observations agree with the concepts presented in your talk and can you accept his proposition that the enzyme ribonucleotide diphosphate reductase plays the key role in the control of growth rate of cells?

G. *Weber*: In an earlier work I compared in a Table the activities of the synthetic and catabolic enzymes of DNA biosynthesis and pyrimidine degradation (G. Weber, S.F. Queener and J.A. Ferdinandus, Advances in Enzyme Regulation 9 (1971) 63). All activities were expressed on the same basis, as picomoles of substrate metabolized per hour per mg protein. In this scale ribonucleotide reductase, with an activity of 3, and DNA polymerase, with an activity of 56, were the lowest among all the enzymes. Thus, the reductase seems to be the rate-limiting enzyme of the de novo synthesis of DNA, and the polymerase that represents the final common pathway of DNA synthesis is the second lowest. This emphasized the importance of these two enzymes. It is relevant that both these enzymes correlate positively with growth rate of the hepatomas. But I consider even more significant the fact that with the increase of the synthetic enzymes the enzymes of pyrimidine catabolism, which exhibit activities orders of magnitude higher than those of the synthetic enzymes, decrease in parallel with tumor growth rate. This reciprocal behavior of the synthetic and catabolic enzymes and their linking with the increase in the expression of malignant properties I view as the most important alterations in this area of intermediary metabolism.

B. *Chance, University of Pennsylvania*: May I raise a point of ignorance? Is there any better correlation in the in vitro and in vivo activities of nucleotide metabolism than of carbohydrate metabolism where we know there is very little correlation. Regulation is the important factor, not in vitro activities.

G. *Weber*: We examined this extensively and observed that the best correlations with tumor growth rate are displayed by the increase in activities of ribonucleotide reductase and DNA polymerase and by the ratios of the incorporation of thymidine into DNA/degradation of thymidine to CO_2. There are also close relationships between tumor growth rate and the increase in the ratios of the key glycolytic/key gluconeogenic enzymes. Now there are over 70 biochemical parameters that correlate with growth rate of the tumors (G. Weber, Ordered and Specific Pattern of Gene Expression in Differentiating and in Neoplastic Cells. In: Differentiation and Control of Malignancy of Tumor Cells, 151-180 (W. Nakahara, T. Ono, T. Sugimura, and H. Sugano, eds.), Proc. 4th International Symposium of the Princess Takamatsu Cancer Research Fund, University of Tokyo Press, Tokyo (1974).

J. Buchanan, Massachusetts Institute of Technology: I would like to add that the enzyme, ribonucleotide reductase, is essentially absent in unfertilized arbacia eggs. After fertilization the reductase is formed de novo and probably is the limiting enzyme in the anabolic reactions of DNA synthesis and hence cell division. So this is another example of the pivotal position of ribonucleotide reductase.

G. Weber: The reductase is undoubtedly an important enzyme and this fact is also brought out by the successful design of anti-cancer chemotherapeutic drugs that are directed against this enzyme activity in animal tumors and in neoplasms in man.

S. Grossman, Union Carbide Corporation: What is the status of these transformation-linked enzymes in precancerous nodules?

G. Weber: We do not have enough data to provide information on this biological material as yet.

OXYGEN REDUCTION BY CYTOCHROME OXIDASE A POSSIBLE SOURCE OF CARCINOGENIC RADICAL INTERMEDIATES

BRITTON CHANCE
Johnson Research Foundation
Department of Biochemistry and Biophysics
University of Pennsylvania School of Medicine
Philadelphia, PA. 19174

Abstract: The increasing awareness of the potential carcinogenic action of radical intermediates of oxygen reduction calls for a special consideration of the cytochrome oxidase pathway; what are the intermediates of oxygen reduction, how are they formed, are they released to the external phase or are they maintained in a special "sequestered" state? The results of low temperature experimentation ($\sim$-100°) on the reaction of cytochrome oxidase with oxygen affords evidence on the nature of "oxy" and "peroxy" intermediates of oxygen reduction by cytochrome oxidase. These intermediates appear to be tightly bound to the oxidase and to exhibit rapid interconversions even at very low temperatures. They afford examples of how the principal pathway of oxygen utilization in the cell employs "sequestered" intermediates rather than those in which reactive radical intermediates are allowed to diffuse through the cell with possible carcinogenic consequences.

GENERAL CONSIDERATIONS

A theme for the bioenergetics section of this symposium is the role of O_2 reduction intermediates in carcinogenesis (anion radicals, peroxides, and epoxides) on enzyme-directed cytotoxicity. Generally, our knowledge of the natural processes limits our understanding of the carcinogenic process and much important work on the normal tissues is still required to identify the steps on the pathway to cancer (1). Recent work on how cytochrome oxidase reduces oxygen in one or two electron steps (2) suggests that oxygen is reduced and bound to the oxidase without the generation of free radical intermediates in solution. The several intermediates of O_2 reduction show that their kinetics and concentrations are related to the energized or de-energized states of native, membrane-bound cytochrome oxidase. In this respect it is useful to compare hemoglobin and cytochrome oxidase, in oxyhemoglobin

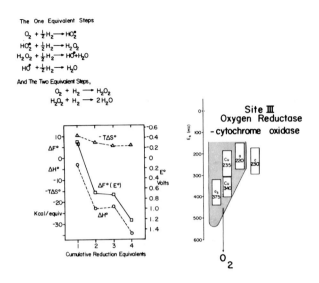

Figure 1. Thermodynamics, chemistry and catalysis of oxygen reduction. The chemical equations indicate one and two step reduction mechanisms with the diagram showing the corresponding thermodynamic properties of these pathways. The inset on the right indicates the components of cytochrome oxidase essential to oxygen reduction (Free energy data according to P. George, in: Oxidases and Related Redox Systems (T.E. King, H.S. Mason, and M. Morrison, eds.) John Wiley & Sons, New York, 1965, p. 3). (MD-443)

there is minimal electron transfer to oxygen, while cytochrome oxidase "throws the book" of electrons at the oxygen molecule (3).

The energy gaps from O_2 to water are principally from O_2 to H_2O_2 and H_2O_2 to H_2O, thus a preferred two-electron, two-step O_2 reduction is preferable to 4 one-electron steps. These free energy changes energize the site III energy coupling in the span from 230 mV at the level of cytochrome c to the level of the O_2 - H_2O couple at $\sim$+800 mV.

Current knowledge about the structure and location of cytochrome oxidase and cytochrome c as membrane proteins is afforded by Figure 2. Cytochrome c is not, strictly-speaking, a membrane protein, it has no hydrophobic tail like cytochrome b_5 (4) and is thus capable of more rapid lateral diffusion in the place of the lipid membrane, interdigitating between cytochrome c_1 the electron donor and cytochrome a, the electron acceptor.

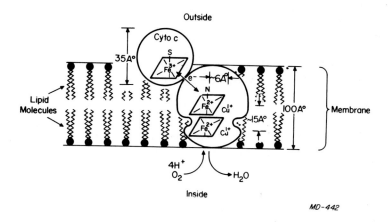

Figure 2. A hypothetical structure for cytochrome oxidase emphasizing its properties as a transmembrane protein and showing cytochrome c as electron donor. The size of cytochrome c is correct while the sizes for cytochrome oxidase are based upon estimates of molecular weight; the structure of cytochrome oxidase is still to be determined.

A variety of studies indicate cytochrome oxidase to be deep in the membrane and that it is a transmembrane protein (5). As shown by Wikström, cytochrome oxidase is sensitive to transmembrane diffusion potentials. Low dose electron micrographs of cytochrome oxidase in monolayers may ultimately give its structure as in the case of bacterial rhodopsin (R. Henderson and J.S. Leigh, Jr., personal communication). Cytochrome oxidase may be tentatively thought of as a dimeric molecule like hemoglobin but with 2 of the hemes replaced by copper atoms. It may have the hemes and coppers within 6°A of one another, at least in certain orientations.

Now let us consider the function of cytochrome oxidase in O_2 reduction, initially in the absence of energy coupling reactions and describe our identification of intermediates by optical and EPR spectroscopy. While cytochrome oxidase might have been considered most likely to form oxy compound analogous to its CO compound and additionally to form a series of oxygen reduction intermediates, up to a year ago, only one "oxygenated" compound of unknown composition had been detected (6). Neither has Gibson (7) found any evidence for functional intermediates in the kinetics of the cytochrome oxidase oxygen reaction at room temperatures at times of several μsec and concluded that the classical "oxygenated intermediate" was non-functional.

EXPERIMENTAL

Since we had already demonstrated light-induced electron transfer from cytochrome $\underline{c}$ to the light-activated reaction center of photosynthetic bacteria in the frozen state and down to 4°K (8), it seemed to me that we should try the same technique in the cytochrome oxidase oxygen reaction. This is readily done by ligand replacement of CO for O_2 by laser flash photolysis of the $\underline{a_3 \cdot CO}$ compound as in Gibson's studies (6) and ours (9) but we have greatly modified the procedure to permit studies in frozen suspensions of mitochondria (10). We have more recently devised a simple and effective method of obtaining optical and EPR difference spectra of such intermediates.

This method has been briefly described (cf. ref. 11, p. 87) and is uniquely suited to the study of differential spectral changes in frozen suspensions of mitochondria, particularly those that are to be examined by both optical and EPR techniques. The problem of light transmission through the EPR tube has been solved by an effective light guide coupling which operates through the walls of an EPR dewar as well (10). The block diagram of the system and an example of its application to these changes has been afforded previously (Figures 5 and 6). In principle the method takes advantage of the fact that the frozen contents of the EPR tube will not exhibit thermal mixing across boundaries established by light-induced activation of the oxygen reaction. Thus the split-beam split photolysis method establishes a baseline in which any inhomogeneities of the contents of the EPR tube between the top half and the bottom half are corrected for in a simple memory circuit and an appropriate gain modulator. Half the tube is laser-illuminated. According to the diagram at the right-hand side of Figure 3, that portion of the EPR tube will then exhibit photolysis and reaction of the contents to form Compound A with reference to the unilluminated, unreacted portion of the tube. After 5 minutes at -110°, Compound A is converted to Compound B. Once Compound B is formed, the unilluminated portion of the tube is activated

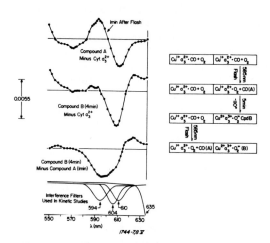

Figure 3. The use of "split-beam, split-photolysis" technique for delineating the absorption difference spectra of Compound A ("oxycytochrome oxidase") and Compound B ("peroxycytochrome oxidase"). The block diagram on the right indicates the successive steps in illumination of the two portions of the EPR tube corresponding to the time-shared light beam of the split beam spectrophotometer. The spectra on the left (Compound A and Compound B) are corrected for the absorption of the reference compound $a_3^{2+} \cdot CO$. The third spectrum from the top, Compound B, is obtained directly from the photolysis apparatus as a difference between the early and late spectra attributed to Compounds A and B.

---

by a laser flash to form Compound A. The difference spectrum of the two portions of the tube represents the difference between Compound A and Compound B shown in the diagram. For kinetic studies, interference filters are employed to observe the kinetics of cytochrome responses. On the bottom trace, the spectra of these filters are recorded with this same apparatus together with a portion of the spectrum of the reference filter at 635 nm.(See Fig. 6 for typical recordings)

The spectra shown in the left-hand side of the figure represent Compound A, top; Compound, B, second; and B-A, third. For convenience, the first two spectra are referenced to that of the reduced cytochrome oxidase in a separate experiment without oxygen in the tube and with flash photolysis at such a low temperature that CO does not recombine (-130°). The spectrum of the carbon monoxide compound is subtracted from that of the first two spectra. The third spectrum is obtained directly from the split-beam split photolysis apparatus.

At -98° we find two intermediates involved in a precursor-product relationship, similar to the sequence of Compound I and II in peroxidases. However, the first compound is an oxy compound (A) of optical and EPR properties very similar to those of the CO compound, but, of course, the kinetic, equilibrium and photo-sensitivities vastly differ. In a few minutes at $-98°$, the oxy compound is converted to a compound (B) that has lost a portion ($\sim1/2$) of the characteristic absorption of reduced heme, and, as optical data show, acquires absorption of partly oxidized copper suggesting a reaction product that contains about half of the heme and copper in the oxidized states. This is also vividly demonstrated by EPR assays of the reaction product which are shown in 3 stages in Fig. 4. When the oxy compound is formed,

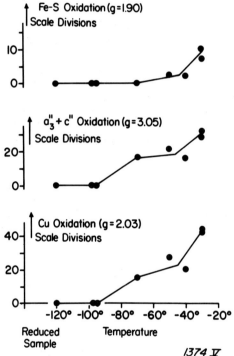

Figure 4. EPR derivative signals obtained in temperature experiments corresponding to Compound A ("oxycytochrome oxidase") at -120 to $-100°$, Compound B ("peroxycytochrome oxidase") at -100 to $-60°$, and to complete oxidation of the components of cytochrome oxidase and the associated iron sulfur protein as well, -60 to $-40°$. Samples optically monitored in EPR tube according to the previous figure are trapped in the progress of the interconversion of the intermediates and are measured by EPR at the g values indicated in the figure. The amplitudes of the differentiated signal are indicated on the abscissae. The EPR signals are measured at low temperatures appropriate to particular components.

---

no heme or copper oxidation occurs ($-100°$), when the conversion to peroxy compound occurs ($-80°$), half of the total oxidized copper and heme signals appear, and finally ($-50°$), the iron sulfur signal appears as all the cytochromes are oxidized and oxygen is reduced to water.

94

A tentative structure for the peroxy compound is one in which peroxide is bridged between copper and heme, based upon the primary interaction of oxygen with heme and the subsequent oxidation of copper and heme. The relationship between the two hemes and coppers which may be more intimate than our diagram indicates and electron exchange between the two parts of the dimer may occur even at low temperatures. Nevertheless, we currently represent the product at low temperature (Compound B), as oxidation of only half of the oxidase (two electron equivalents, one from heme and one from copper) and consequently contains half-reduced oxygen.

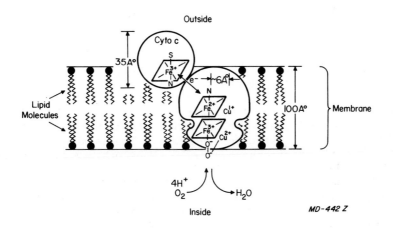

Figure 5. A hypothetical configuration of oxygen in relation to iron and copper appropriate to Compound B ("peroxycytochrome oxidase"). The schematic diagram is consistent with the lack of reactivity of the unliganded form towards peroxide and the lack of inhibition of cytochrome oxidase activity by catalase.

The kinetics of interconversion of the intermediates A and B are plotted in Fig. 6, according to kinetic studies at the wavelengths near those indicated at the bottom of Fig. 3 for heme kinetics. Copper kinetics are recorded at 830-940 nm. Compound A, oxycytochrome oxidase, rises rapidly and falls to zero as would be expected of a primary intermediate, such as the peroxidase compound I, followed by product curves at the heme and copper wavelengths due to the formation of the peroxy product. The kinetics are similar to the precursor-product curves of the peroxidase compounds I and II.

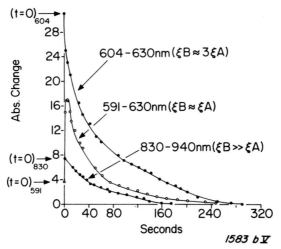

Figure 6. Kinetics of formation and interconversion of oxygen reduction products of cytochrome oxidase measured at $\sim -90°$. The trace at 591 nm represents mainly the formation and disappearance of Compound A while the traces at 604 and 830 nm represent respectively the oxidation of the heme and copper components of the cytochrome oxidase. Traces measured by multi-channel spectrophotometer, ~ 5 μM cytochrome oxidase in 15 mg/ml beef heart mitochondria supplemented with 10 mM succinate, 10 mM glutamate, saturated 1.2 mM carbon monoxide in a medium containing 40% ethylene glycol, 60% 0.2 M mannitol sucrose with 20 mM phosphate present, oxygenated at $-20°$ with 300 mM oxygen, flash photolyzed at $-90°$.

The kinetics characteristically show the rapid formation of the oxy compound, Compound A, but at a low steady state concentration. The kinetics of this reaction are influenced by ATP, the reaction being several times more rapid when a configuration of the oxidase established by pre-treatment of the mitochondria by ATP at room temperature is used in these experiments (11). The dissociation constant for oxygen is large and high oxygen concentrations are required to reach saturation; the saturation of cytochrome oxidase with oxygen at the steady state point in these kinetics is only 30%. Two reaction products are observed at wavelengths appropriate for the heme oxidation (604 nm) or for copper oxidation (830 nm). At 830 nm, there is no interference from Compound A, and the trace appears to start monotonically and to reach the maximal rate when the concentration of Compound A is maximal. At 604 nm, the majority of the absorption is due to heme oxidation but a small initial portion is due to interference with

Compound A kinetics. Nevertheless the precursor-product sequence of Compound A and oxidized heme and copper appears generally to be substantiated by these kinetics. The kinetics of disappearance of Compounds A and B are strongly influenced by the ATP concentration, again being more rapid in the ATP-treated membranes.

Detailed studies are, however, under way to determine, by optimized computer fits of the differential equations representing sequential reactions; the minimum number of intermediates involved in these kinetics. At present it appears that the crucial question of whether copper and heme donate electrons simultaneously to reduce oxygen to water is susceptible to analysis by these methods.

DISCUSSION AND SUMMARY

In summary, we find both the heme and copper moieties of cytochrome oxidase are active in electron transfer to oxygen and probably among themselves. It is even possible that all four components of the cytochrome oxidase, 2 hemes and 2 coppers, participate in these low temperature electron transfer reactions, especially by tunneling mechanisms as demonstrated in photosynthetic systems (8). The relevance to radical intermediates in carcinogenesis is that the major pathway of oxygen reduction in the cell exhibits an initial binding of O_2 following which all steps are highly irreversible and involve tightly bound species of O_2 reduction products. Cytochrome oxidase reduces oxygen in "sequestered reactions" where the intermediates are not in equilibrium with the external phase. In this way, we are protected from free radical intermediates that might be quite damaging in view of the large metabolic function of the major pathway of cell respiration.

REFERENCES

(1) E. Racker, in: MTP Intern. Review of Science, I, Vol. 3, ed. E. Racker (Butterworth and University Park Press, London and Baltimore, 1975) p. 163.

(2) B. Chance, C. Saronio, and J.S. Leigh, Jr., Proc. Natl. Acad. Sci., 72 (1975) 1635.

(3) B. Chance, C. Saronio, and J.S. Leigh, Jr., J. Biol. Chem., 250 (1975) 9226.

(4) A. Tsugita, M. Kobayashi, T. Kajihara, and B. Hagihara, J. Biochem., 64 (1968) 727.

(5) M.K.F. Wikström, in: Electron Transfer Chains and Oxidative Phosphorylation, eds. E. Quagliariello et al. (North-Holland, Amsterdam, 1975) p. 102.

(6) Y. Orii and K. Okunuki, J. Biochem. (Tokyo), 53 (1963) 489.

(7) C. Greenwood and Q.H. Gibson, J. Biol. Chem., 242 (1967) 1782.

(8) D. DeVault, J.H. Parkes, and B. Chance, Nature, 215 (1967) 642.

(9) M. Erecinska and B. Chance, Arch. Biochem. Biophys., 151 (1972) 304.

(10) B. Chance, N. Graham, and V. Legallais, Analy. Biochem., 67 (1975) 552.

(11) B. Chance, J. Harmon, and M.K.F. Wikström, in: Electron Transfer Chains and Oxidative Phosphorylation, eds. E. Quagliariello et al. (North-Holland, Amsterdam, 1975) p. 81.

Acknowledgment

This research was supported by USPHS HL-15061, GM-12202, HL-17826, and CM-53830.

Discussion

G. Weber, Indiana University: I just have a simple question. In what particular system were your determinations carried out?

B. Chance, University of Pennsylvania: Well we have studied mitochondria from all types of tissues: beef heart, rat heart and liver, yeast, flies, <u>Micrococcus denitrificans</u> and a whole variety, - in other words everything that is a eukaryote or might be a eukaryote precursor. The organelle is the intact functional mitochondrion. The mitochondria can be cycled through this process and brought back to room temperature. They will accumulate ions, make ATP, exhibit respiratory control, exhibiting their physiological properties.

A. Theorell, Karolinska Institutet: Ladies and gentlemen, I am quite sure all of you, like myself, are deeply interested in these techniques that have been refined to a degree never before possible. How small differences in wavelengths which may cause shifts in curves can then be interpreted as part of the whole reaction system. I have had the pleasure of knowing Britton Chance's work since an early time and I suppose all of you know that he started very young (even before World War II) making the most considerable inventions, for instance - how to improve radar. Now he is concentrating on much smaller distances, i.e. dealing with angstroms. But the results are not in proportion to wavelengths but perhaps rather the opposite. The smaller the wavelength the more interesting the results. I am very interested in these procedures, both because they are so accurate and because they are very rapid. I suppose this may have contributed to the fact that Britton Chance has produced a tremendous amount of work during his life. When reactions require at the most milliseconds, then of course you can achieve a great many results in your life. He has really accomplished that. As he indicated to you in the beginning, we had early collaboration since I was interested in rapid reactions of enzymes, coenzymes and so forth, and the obvious possibility was to go to Philadelphia to work on it further with Britton Chance.

I would not raise any specific questions after what we have seen today. But this is a very big field, of course, and I suppose, to take only one little example, one could find out in what respect microorganisms are similar to the tissues of higher animals. The advantage with all this is that you can produce a lot of interesting results in a reasonable time. At this time I believe that the number of Britton Chance's publications has exceeded 1,000. I will not take any more time with this aspect of his interesting work since I am sure there are many in the audience who may need an explanation of the results presented to us by Britton Chance. Thank you.

B. Vallee, Harvard University: I did not quite catch what metal distance you described. Is it the distance from the copper to the porphyrin that is 6 angstroms?

B. Chance: We used EPR for the copper and optical absorption at 830. Does that answer your question?

B. Vallee: Well, not exactly. I wondered which distance you described and by what means it was determined.

B. Chance: Experimental data indicate the two hemes are close together (J.S. Leigh, D.F. Wilson, C.S. Owen, and T.S. King, Arch. Biochem. Biophys. <u>160</u>, 476(1974).

B. Vallee: Thank you.

R. Estabrook, University of Texas: Your presentation is certainly provocative and brings up a number of questions, but I will try and restrict myself to only a few. Is the oxy-intermediate that you described similar to that which Ludwig described a number of years ago in your own laboratory while studying the interaction of oxygen with the reduced cytochromes of yeast?

B. Chance: If we extrapolate from $-100°$ to $+25°$ at 9.9 Kcal for the oxycompound and 12.5 Kcal for the peroxy compound we find the concentration of oxy will be low at $+25°$.

R. Estabrook: Second, as I understand you, you indicated the computer model suggested that step two and step four were committed and essentially irreversible, yet your final scheme showed these reactions to be reversible.

B. *Chance:* Well they are formally reversible, but the backward rates are insignificant.

R. *Estabrook:* Lastly, your abstract indicated that you were going to discuss ATP effects on this reaction. Could you briefly summarize which of the 4 reactions are influenced by ATP or the energy conserving potential?

B. *Chance:* ATP induces a structure of the oxidase which is more open to oxygen, donates electrons to oxygen more effectively, is recalcitrant to accepting electrons from cytochrome a and cytochrome c. This gives support for those old crossover data between a and c, explains Muroaka and Slater too since their crossover points have to do with intermediates they were unaware of. So the structure of energized cytochrome oxidase is vastly different, the rates are different in the ATP soaked membranes as compared with uncoupler soaked membranes, as examined at low temperatures. There is a real structure-function coupling in cytochrome oxidase.

N. *Kaplan, University of California:* That first intermediate is a very exciting one. Do you think it has any relationship to the pathway in which oxyhemoglobin is formed? Is it possible to study the mechanism of formation of the intermediate?

B. *Chance:* There is a parallelism between oxyhemoglobin and oxycytochrome oxidase. We can estimate highly dissociated oxy compound, the equilibrium constant at low temperature is probably diffusion controlled on the "on" side and thermally controlled on the "off" side. The dissociation constant will increase at room temperature, so the trick in the oxidase is that whenever an oxygen molecule reacts with iron it is reduced before it can get out.

H. *Kareem, University of Miami:* My comment is on the superoxide dismutase enzyme. Is this enzyme involved in the proposed scheme? Otherwise how will the conversion of copper-two-plus-oxygen-two-minus complex i.e. Cu^{2+} a_3^{3+} $O_2^=$ to the next product be achieved with your proposed mechanism?

B. *Chance:* Superoxide dismutase does not effect the activity so the superoxide anion seems to stick onto the iron. We have looked at EPR spectra of these intermediates to see whether we get the O_2^- radical. We have not, but we also have not maximized the concentration of intermediate II.

A. *Mildvan, Institute for Cancer Research:* If I could just comment very briefly on that small point. Your intermediate II is Cu^{2+} $O_2^-\cdot$ so you have a superoxide directly coordinated to a paramagnetic copper and this should be very difficult to observe by electron spin resonance. It might be immeasurably broad. Perhaps some other magnetic technique might be required to see that. One small question. You never said what step 4 was on that very nice computer fit. I assume that was the second electron getting from the copper to consumate the entire process making O_2^{2+} and Cu^{2+}.

B. *Chance:* That is correct. The intermediate 4 is consistent with the two electron reduction intermediate. The first point is a very good one and I should have thought of it myself. We would get paramagnetic broadening. That leaves us doing volume magnetic susceptibility I guess which is usually where you end up in the study of most of these compounds.

THE ACTIVATION OF POLYCYCLIC HYDROCARBONS: CYTOCHROME'S P-450, OXYGEN AND ELECTRONS

RONALD W. ESTABROOK, VIRGINIA W. PATRIZI, and RUSSELL PROUGH
Department of Biochemistry, The University of Texas
Health Science Center at Dallas, Dallas, Texas 75235

Abstract: The oxidative metabolism of polycyclic hydrocarbons leads to products of high cellular toxicity or mutagenicity presumed responsible for the expression of cellular proliferation and tumorigenesis. A unique electron transport system, in which cytochrome P-450 serves as the terminal oxidase, is responsible for oxygen activation and insertion into the substrate molecule. Methodology has been developed to characterize the variety of metabolites formed. The specificity of site of oxidation of complex organic molecules, such as benzo(a)pyrene, is now amenable to study. The nature of "active oxygen", postulated as an oxene or perhydroxy anion, may dictate the pattern of metabolites resulting from the cytochrome P-450 catalyzed conversion of benzo(a)pyrene.

Recently reports have appeared in the popular press indicating that as great as 80 percent of cancer may be attributed to chemical carcinogenesis. It is common knowledge that chemicals, for example, polycyclic aromatic hydrocarbons present in tobacco smoke, have the potential of initiating tumorigenesis - a topic of great public concern as evidenced by federal and state legislation resulting subsequent to the report in 1964 of the Surgeon-General on "Smoking and Health" (1). Since the recognition in the 1930's (2) that topical application of chemicals such as benzo(a)pyrene to the skin of mice results in malignancy, the biomedical research community has pursued various courses of experimentation to understand the underlying mechanisms operative in this type of chemical carcinogenesis. The current hypothesis (3) which serves as the framework for a wide variety of studies states that polycyclic hydrocarbons are oxidatively converted from a precarcinogen to a carcinogenic active metabolite; this metabolite (or metabolites) can then interact with nucleic acids or proteins thereby altering the genetic and homeostatic characteristics of the cell. The purpose of the present paper is to briefly summarize our current knowledge concerning the initial step in this process, i.e. the oxidative transformation of polycyclic hydrocarbons to a variety of metabolites.

Mixed Function Oxidation Reactions - Most tissues of mammals contain an electron transport complex, generally associated with the endoplasmic reticulum, which functions in the oxidative transformation of a wide variety of both endogenous and exogenous complex organic compounds. Concomitant with the oxidation of reduced pyridine nucleotide (Figure 1), molecular oxygen is "activated" for insertion into the substrate rendering it more polar, i.e. more water soluble, permitting its subsequent conjugation for transport and excretion.

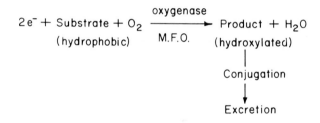

Figure 1

The types of substrates metabolized in this manner encompass a wide diversity of organic compounds ranging from steroids, drugs, insecticides and anesthetics to polycyclic hydrocarbons. In the case of many of these compounds, such as a large number of drugs, this enzymatic conversion is considered a detoxification reaction (Fig. 2). Indeed, the duration of drug effectiveness is frequently linked to how rapidly it is oxidatively transformed (4). In contrast, as indicated above, with compounds such as polycyclic hydrocarbons, the product formed is considered to be of significantly higher toxicity than the original substrate. Thus, the dualism of this enzyme system may be looked upon as both beneficial or detrimental to the organism.

A. Active Drug $\xrightarrow{O_2, e^-}$ Inactive Drug

B. Precarcinogen $\xrightarrow{O_2, e^-}$ Carcinogen

Figure 2

One of the keys to understanding the molecular basis of this type of mixed function oxidation reaction resides with the elucidation of the mechanism of function of a class of hemoproteins, termed cytochromes P-450. In mammalian tissues cytochrome P-450 is a membrane bound pigment functioning as the terminal electron transport oxidase whereby molecular oxygen is activated and bound substrate molecules are oxidized. Cytochrome P-450 apparently exists in multiple forms (5,6) which can be preferentially induced upon pretreatment of animals with a number of different types of substrates. For example (Figure 3), introduction of phenobarbital either intraperitoneally or in the drinking water of rats results in a rather rapid three- to four-fold increase in the liver content of one form of cytochrome P-450 (7). A single injection of a polycyclic hydrocarbon, benzo(a)pyrene or 3-methylcholanthrene, causes a two- to three-fold increase in the liver content of this pigment, in this case a modified cytochrome P-450, termed P-448 (8).

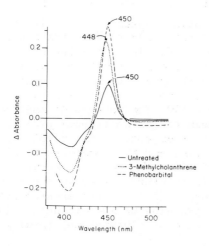

Figure 3
The effect of animal pretreatment with phenobarbital (--) or 3-methylcholanthrene (···) on the content of cytochrome P-450 (P-448) in liver microsomes.

These various types of cytochrome P-450 differ in their electrophoretic mobility (5), interaction with substrates (9), and ability to oxidatively metabolize substrates (10). Thus, there appears to be a limited range of substrate specificity dictated by the type of cytochrome P-450 associated with the microsomal electron transport system.

The Metabolism of Polycyclic Hydrocarbons - An abbreviated scheme illustrating the oxidative conversion of polycyclic hydrocarbons as catalyzed by cytochrome P-450 (P-448) is shown in Figure 4.

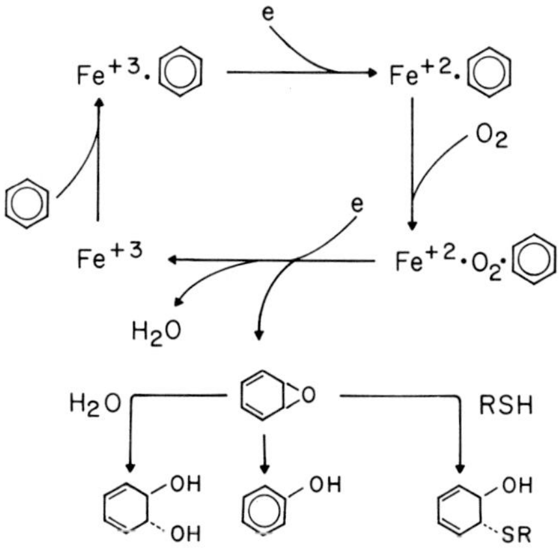

Figure 4
Schematic of oxidative metabolism of polycyclic hydrocarbons.

Analogous to studies with other substrates (11) it is presumed that chemicals such as benzo(a)pyrene interact with the ferric form of the hemeprotein forming a complex which can undergo enzymatic reduction. The resultant ferrous substrate complex then interacts with oxygen to form a ternary complex consisting of substrate, oxygen and the ferrous hemeprotein: termed oxy-cytochrome P-450. The subsequent addition of an electron, ultimately originating from reduced pyridine nucleotide and transfered via a flavoprotein, may result in the formation of an epoxide, water, and the regeneration of the ferric form of cytochrome P-450. The proposed epoxide can undergo enzymatic hydrolysis to dihydrodiols in the presence of the enzyme epoxide hydrase (12); or the epoxide can undergo nucleophilic attack by proteins or nucleic acids (13) (illustrated in Fig. 4 by RSH, such as glutathione) forming a variety of different products some of which may be detrimental to the cell. Alternatively, the epoxide can undergo rearrangement to form phenols or phenols may be formed directly by ring hydroxylation without the need for an epoxide intermediate. Critical to our understanding of the carcinogenic effects of epoxides or other reactive intermediates is a better understanding of the cyclic function of cytochrome P-450, in particular, the nature of "active oxygen", and the pattern of products formed during the metabolism of such polycyclic hydrocarbons.

Fluorometric Analysis of 3-Hydroxybenzo(a)pyrene - The extensive resonant conjugation present in polycyclic hydrocarbons affords the opportunity to apply fluorescence analysis for the measurement of enzymatic activity (14,15). Indeed, the sensitivity of this technique has been used for a number of years to survey the rate of formation of one of the products, 3-hydroxybenzo(a)pyrene, during the metabolism of benzo(a)pyrene by a variety of tissues subjected to a number of different metabolic perturbations (16). As illustrated in Fig. 5, extracts of tissue incubations can be analyzed at various time points using 467 nm as an excitation wavelength and 515 nm as the measuring (emission) wavelength for the determination of 3-hydroxybenzo(a)pyrene.

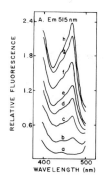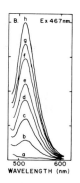

Figure 5
Fluorescence spectral analysis of the oxidative metabolism of benzo(a)pyrene with liver microsomal fractions from 3-methylcholanthrene-treated rats. The time course of metabolism was determined at the following time points: a. 0 min, b. 0.5 min, c. 1.0 min, d. 1.5 min, e. 2.0 min, f. 3.0 min, g. 4.0 min, and h. 5.0 min. The time course was performed using a modification of the method of Nebert and Gelboin (15).

Using this approach, patterns of the time course of 3-hydroxybenzo(a)pyrene formation can be determined as shown in Fig. 6. The effect of pretreatment of an animal with a pre-carcinogen, such as 3-methylcholanthrene, resulting in the induction of cytochrome P-448 in liver or lung, can be directly evaluated. As illustrated in Fig. 6 the influence of other substrates of cytochrome P-450, such as the drug ethylmorphine, can be shown to modify the rate of 3-hydroxybenzo-(a)pyrene formation. Contrary to many reports in the literature (14,15), our studies show that the rates of formation of 3-hydroxybenzo(a)pyrene are not linear with time; a progressive decrease in specific activity was observed with increasing time of incubation. This non-linearity of the reaction could result from the destruction of the enzyme, the development of product inhibition, or the further metabolism of 3-hydroxybenzo(a)pyrene to metabolites not detected by the fluorometric method of analysis. Further, the ability to monitor the formation on only 3-hydroxybenzo(a)pyrene is rather tedious and restrictive and provides little or no information about other products formed during the reaction.

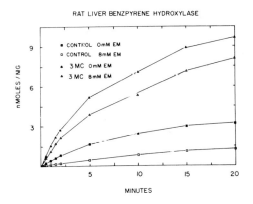

Figure 6
Time course of formation of fluorescence benzo(a)pyrene phenols with liver microsomal fractions from untreated (□) and 3-methylcholanthrene (△) pretreated rats. Closed symbols indicate time course experiments performed in the absence of ethylmorphine and open symbols indicate time course experiments performed in the presence of 8 mM ethylmorphine. The experiments were performed using a modification of the method of Nebert and Gelboin (15).

High Pressure Liquid Chromatographic Analysis - Major advancements in the analysis of the multiple products formed during benzo(a)pyrene metabolism occurred a few years ago when Selkirk et al. (17,18) published the application of high pressure liquid chromatography for the resolution of phenols, dihydrodiols, and quinones. This technique coupled with the elegant skill of Jerina et al. (19,20) for the chemical synthesis of many of these products, now provides us the base for more elaborate metabolic studies. For example (Figs. 7 and 8) marked differences in the pattern of products formed and the ratio of various products can be readily demonstrated when using microsomes from liver and lung (Fig. 7); changes in the relative amounts of quinones and various phenols are apparent. Likewise, dramatic differences are observed when comparing products formed (Fig. 8) when microsomes from liver of untreated, phenobarbital and 3-methylcholanthrene treated animals are employed. The latter experiments are of potential significance because of the known susceptibility of lung tissue to the development of cancer.

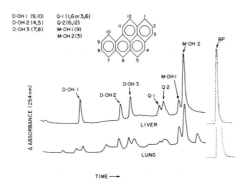

Figure 7
High pressure liquid chromatography separation of benzo(a)pyrene metabolites formed in incubation mixtures containing microsomal fractions of liver and lung from 3-methylcholanthrene pretreated rats. Chromatographic separation was performed as described (18,19). Products were monitored by their absorption at 254 nm. The reaction mixture contained 80 μm benzo(a)pyrene, 0.05M Tris-HCl buffer, pH 7.4, 0.3 mg/ml liver microsomal protein, 1mM NADPH, 5 mM $MgCl_2$, 5 μM $MnCl_2$ 5mM isocitrate, 0.1 u/ml isocitrate dehydrogenase at 37°C.

RATE OF FORMATION OF BENZO(a)PYRENE METABOLITES

Treatment	Total	9,10 diOH	4,5 diOH	7,8 diOH	Quinones	Phenol 1	Phenol 2
Control	3.1	0.2	0.09	0.07	0.8	0.25	1.7
Phenobarbital	1.65	0.2	0.2	0.06	0.65	0.04	0.5
3-MC	4.2	0.5	0.25	0.30	0.8	0.5	1.8

Rat Liver Microsomes; Rates as nmole product/min/nmole P-450

Figure 8
The analysis of the rate of metabolite formation from benzo(a)pyrene oxidation in incubation mixtures of liver microsomal fractions from untreated, phenobarbital treated or 3-methylcholanthrene-treated rats. ^{14}C-labelled metabolites were separated by H.P.L.C. and quantitated by liquid scintillation spectrometry (18,19) using the conditions of Fig. 7.

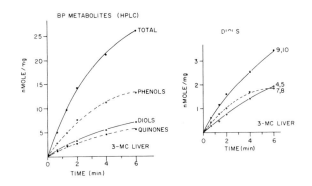

Figure 9
Time course of metabolism of benzo(a)pyrene catalyzed by liver microsomal fractions from 3-methylcholanthrene treated rats. Analysis was performed as described (19) using the conditions cited in Fig. 7. The left hand figure shows the total metabolism and rate of formation of the major metabolites of benzo(a)pyrene. The right hand figure shows the rate of formation of individual dihydrodiols.

Applying this technique, studies are underway to characterize the time course of product formation (Fig. 9) so as to permit an evaluation of the basic enzymology directing the oxidative metabolism of benzo(a)pyrene, as well as, the ability to recognize, isolate, accumulate, and then characterize as yet unknown new metabolites. The ultimate goal of these studies is to establish the relative potential carcinogenicity or mutagenicity of the various metabolites and then to develop means to direct metabolism toward non-toxic products that can be rapidly excreted from the organism.

Spectrophotometric Studies During Benzo(a)pyrene Metabolism - Much of our knowledge of cytochrome P-450 function has come from a detailed spectrophotometric analysis of oxidation-reduction changes occurring during the aerobic steady-state of substrate metabolism. Recently we have turned our attention to an examination of optical spectral changes associated with benzo(a)pyrene metabolism. We had previously avoided these studies because of the recognized complexity resulting from the absorbance of benzo(a)pyrene itself together with the absorbance bands of reduced cytochromes and the inherent difficulties associated with the study of membrane bound enzymes superimposed on a large background absorbance of substrate. Although the studies are still preliminary, the approach looks very promising as shown in Figs. 10, 11, and 12. Phenols formed during metabolism are each characterized by unique spectral absorbance bands. The application of the repetitive scan technique allows a direct demonstration of differences in metabolite formation as a function of time. Marked changes in the formation of metabolites (as determined by spectrophotometric analysis) using liver microsomes from 3-methylcholanthrene (Fig. 10) and phenobarbital (Fig. 12) treated animals and the influence of the inhibitor (12) of epoxide hydrase, (TCPO, 1,1,1-trichloropropene-2,3-oxide) (Figure 11) are readily discernible. Further evaluation of this approach to the study of benzo(a)pyrene metabolism offers new and exciting possibilities for the study of molecular events associated with the oxidative conversion of this pre-carcinogen.

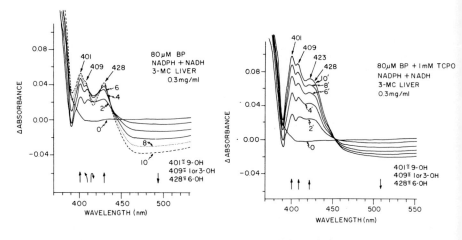

Figure 10   Figure 11

Figure 10
Optical spectra repetitive scan analysis of benzo(a)pyrene phenol formation catalyzed by liver microsomal fractions from 3-methylcholanthrene treated rats. The reaction mixture was that cited in Fig. 7 except the reaction was initiated with addition of NADPH (100 μM) to the sample cuvette and the repetitive scans were collected every two minutes after initiation. The reference and sample cuvettes contained 100 μM NADH to remove the absorbance of cytochrome b_5 from the spectra.

Figure 11
Optical spectra repetitive scan analysis of benzo(a)pyrene phenol formation in the presence of 1,1,1-trichloropropene-2,3-oxide (2mM) catalyzed by liver microsomal fractions from 3-methylcholanthrene treated rats. Analysis was performed as described in Fig. 10.

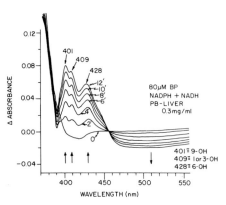

Figure 12
Optical spectra repetitive scan analysis of benzo(a)pyrene phenol formation catalyzed by liver microsomal fractions from phenobarbital treated rats. Analysis was performed as described in Fig. 10.

The Mechanism of Cytochrome P-450 Function - Our knowledge
of the chemistry of cytochrome P-450 and its interaction with
a variety of ligands has increased markedly in the last few
years. Likewise, our understanding of the cyclic function of
cytochrome P-450 during substrate oxidation has expanded
although much more has yet to be learned. The current hypothesis (21) of oxidation-reduction reactions occurring is
summarized in Fig. 13.

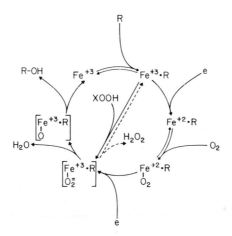

Figure 13

During the aerobic steady-state where an excess of substrate molecules, oxygen, and electrons, ultimately derived
from reduced pyridine nucleotide, the various steps of cytochrome P-450 can be envisioned as followed:

a) The low spin form of the ferric cytochrome P-450
combines reversibly with a substrate molecule (R) to be oxidatively converted resulting in a transition to the high
spin form of ferric cytochrome P-450 complexed with the substrate molecule;

b) The complex of substrate and ferric cytochrome P-450
undergoes a one electron reduction to form the ferrous hemo-
protein-substrate complex;

c) Molecular oxygen reacts with the ferrous hemoprotein-
substrate complex resulting in a ternary complex of substrate,
oxygen, and ferrous cytochrome P-450;

d) This ternary complex undergoes a second one electron
reduction forming a proposed complex containing substrate,
a peroxide anion with the ferric heme or a perferryl form
of cytochrome P-450 with reduced oxygen;

e) Further it is proposed that oxygen is activated and water is expelled from the complex leaving an oxene associated with the heme iron. Electron rearrangement results in the introduction of the oxene into the substrate molecule. The details of the mechanism of oxygen insertion into the substrate molecule and the nature of the electron rearrangement reactions are not known, although the formation of an epoxide is certainly one logical consequence;

f) The resulting oxidized substrate (product) dissociates from the ferric hemoprotein regenerating the low spin form of ferric cytochrome P-450.

Of interest are recent studies which demonstrate that:

g) In the presence of an organic peroxide and absence of reduced pyridine nucleotide the complex of substrate with ferric cytochrome P-450 can be converted directly to a ternary complex comparable to that described in (d). Release of the alcohol form of the organic peroxide and formation of an oxene associated with the heme iron is proposed in a manner similar to the reactions described in (e);

h) In a reaction mechanism, not now understood, the ternary complex of the activated oxygen with cytochrome P-450 may breakdown resulting in the formation of hydrogen peroxide and the regeneration of ferric cytochrome P-450.

Conclusions: Critical to the understanding of the metabolism of polycyclic hydrocarbons and the carcinogenic effects of some of the metabolites formed is a means of predicting and regulating the site of oxygen insertion into these complex organic molecules. Studies with a variety of steroids have suggested (22) the need for steering groups on the substrate to provide an "anchor point" for orientation near the heme and "activated oxygen". Compounds such as benzo(a)pyrene, however, do not have functional groups that may serve in this capacity; consequently a wide variety of possible sites of interaction may occur (Figure 14). One is tempted to speculate that indiscriminant interaction occurs dictated only by the differences in electron density (perhaps perturbed by resonant interaction with the porphyrin ring of the heme). This postulate appears, however, to be oversimplified as indicated by recent studies reported by Coon et al.

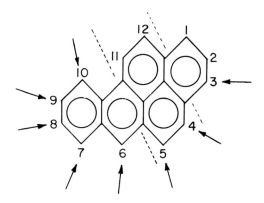

Figure 14
Structure of benzo(a)pyrene with arrows indicating potential sites of oxygen interaction. The dashed area represents the K region of the molecule.

(23) suggesting that unique patterns of metabolites are formed when purified cytochrome P-450 and cytochrome P-448 are employed. Indeed, the studies presented above (Fig. 8) support this concept that specificity can occur. The methodologies now exist for a concerted effort to unravel this puzzle of molecular events. Delineation of the factors directing substrate juxtaposition in its complex with cytochrome P-450, hold the promise of developing means to direct the enzymatic process away from formation of highly toxic and reactive intermediates presumed to be the primary causative agents for chemical carcinogenicity.

ACKNOWLEDGEMENT
Supported in part by research grants from the U.S.P.H.S. GM 16488 and HL 17134 and by a N.C.I. Contract NO1-CP-33362.

REFERENCES

(1) "Smoking and Health. Report of the Advisory Committee to the Surgeon General of the Public Health Service" U.S. Department of Health, Education and Welfare (Public Health Service Publication No. 1103, Washington, D.C. 1964).
(2) J.C. Arcos and M.F. Argus, in: Chemical Induction of Cancer, Vol. IIA (Academic Press, New York, 1974) P. 17.

(3) P. Sims and P.L. Grover, Advances in Cancer Res., 20 (1975) 165.
(4) J.R. Gillette, D.C. Davis, and H. Sasame, Ann. Rev. Pharm., 12 (1972) 57.
(5) D.A. Haugen, T.A. van der Hoeven, and M.J. Coon, J. Biol. Chem., 250 (1975) 3567.
(6) J. Werringloer and R.W. Estabrook, Arch. Biochem. Biophys., 167 (1975) 270.
(7) H. Remmer, H.Greim, J.B. Schenkman, and R.W. Estabrook, Methods Enzymol., 10 (1967) 703.
(8) G.J. Mannering, N.E. Sladek, C.J. Parli, and D.W. Shoeman in: Microsomes and Drug Oxidations, eds. J.R. Gillette, A.H. Conney, G.J. Cosmides, R.W. Estabrook, J.R. Fouts, and G.J. Mannering (Academic Press, New York, 1968) P. 303.
(9) J.B. Schenkman, H. Remmer, and R.W. Estabrook, Mol. Pharmacol., 3 (1969) 113.
(10) A.H. Conney, Pharmacol. Rev., 19 (1967) 317.
(11) J. Peterson, Y. Ishimura, J. Baron, and R.W. Estabrook, In: Oxidases and Related Redox Systems, Vol. 2, eds. T.E. King, H.S. Mason and M. Morrison (University Park Press, Baltimore, 1973) P. 565.
(12) F. Oesch, Xenobiotica 3 (1973) 305.
(13) J.W. Daly, D.M. Jerina and B. Witkop, Experientia 28 (1972) 1129.
(14) L.W. Wattenberg, J.L. Leong, and P.J. Strand, Cancer Res., 22 (1962) 1120.
(15) D.W. Nebert and H.V. Gelboin, J. Biol. Chem., 243 (1968) 6242.
(16) D.W. Nebert and H.V. Gelboin, Arch. Biochem. Biophys. 134 (1969) 76.
(17) J.K. Selkirk, R.G. Croy, and H.V. Gelboin, Science 184 (1974) 169.
(18) J.K. Selkirk, R.G. Croy, P.P. Roller, and H.V. Gelboin, Cancer Res., 34 (1974) 3474.
(19) G. Holder, H. Yagi, P. Dansette, D.M. Jerina, W. Leven, A.Y.H. Lu, and A.H. Conney, Proc. Nat. Acad. Sci., U.S.A. 71 (1974) 4356.
(20) H. Yagi, O. Hernandez, and D.M. Jerina, J. Am. Chem. Soc., 97 (1975) 6881.
(21) R.W. Estabrook, J. Werringloer, E.G. Hrycay, P.J. O'Brien, A.D. Rahimtula and J.A. Peterson, Biochem. Soc. Trans., 3 (1975) 793.

(22) R.W. Estabrook, G. Martinez-Zedillo, S. Young, J. A. Peterson and J. McCarthy, J. Steroid Biochem., 6 (1975) 419.
(23) F.J. Wiebel, J.K. Selkirk, H.V. Gelboin, D.A. Haugen, T.A. van der Hoeven, and M.J. Coon, Proc. Nat. Acad. Sci., U.S.A. 72 (1975) 3917.

Discussion

B. *Chance, University of Pennsylvania*: Do you think there are different P-450's or different attacks by a single P-450 corresponding to each reaction product.

R. *Estabrook, University of Texas*: From the studies that have been carried out to date, all forms of cytochrome P-450 appear capable of hydroxylating benzo (a) pyrene. However the site of oxygen insertion varies from one form of liver P-450 to another. Whether there is a common mechanism of oxygen insertion which is operative is still open for consideration. We are not at that stage of sophistication yet.

S. *Weinhouse, Temple University*: I have a few questions Ron that I would like to ask you. One is could the difference in product pattern between liver and lung be due to the fact that the liver conjugates at least some of the products whereas the lung does not?

R. *Estabrook*: We have not studied the conjugation reaction with our system. Since we are using an isolated, washed, microsomal fraction with none of the necessary substrates for conjugation, and since we are using radioactive benzpyrene we see no evidence of conjugation. In other words, our balance sheet of product recovery is 95%.

S. *Weinhouse*: Now, I would like to ask another general question. In the whole cell, what is the magnitude of the P-450 oxidation in relation to mitochondrial cytochrome oxidase oxidation and is there competition between the two?

R. *Estabrook*: This depends on the organ and the substrate employed. Studies carried out by Thurman and Scholtz in Dr. Chance's lab would suggest that the microsomal system has the potential of utilizing oxygen at a rate about 35% of that obtained with mitochondria. Although you may not have recognized it, the numbers here show that the rate of benzpyrene metabolism i.e. the total metabolites formed, is of the order of about 8 to 10 nanomoles per minute per mg microsomal protein. This means that the rate of benzo(a)-pyrene metabolism is about the same as that seen with other drug substrates. The microsomal oxidation system can represent a significant portion of oxygen utilized.

A. *Mildvan, Institute for Cancer Research*: That was a very nice mechanism that you proposed. I am sure it is consistent with the work of many others. I just wanted to say that from the point of view of possible chemotherapy, do you see any entry into that mechanism for inhibiting the reaction? Alternatively, aside from the obvious approach of eliminating pollution, is it feasible to activate the epoxide in a manner analogous to the function of superoxide dismutase?

R. *Estabrook*: The epoxide hydrase is induced concomitant with the cytochrome P-450 system. This has led then to a basic philosophical argument concerning the possible beneficial effects resulting from inducing the monoxygenase system because you likewise induce the epoxide hydrase. The question is how could you get more epoxide hydrase at the same time you get less P-450. One thing that I did not mention is that no one knows (to my knowledge) how these agents act as "inducers". So many different compounds alter the genic apparatus that we do not fully understand what's going on.

B. *Vallee, Harvard University*: Do you know whether or not similar induction might occur in microbiological systems?

R. *Estabrook*: Is there similar induction in microbiological systems?

B. *Vallee*: Right, for these types of compounds?

R. *Estabrook*: Some very beautiful work has been carried out in the laboratory of Gunsalus showing that the terpene, camphor, is an excellent inducer of a type of cytochrome P-450 in Pseudomonas putida. To my knowledge polycyclic hydrocarbons have not been demonstrated as inducers in microorganisms.

B. *Vallee*: You may have misunderstood. I wondered whether or not the effect of carcinogens of the type you discussed on P-450 have been examined?

R. Estabrook: I do not know. This would be difficult to test because of the difficulty in maintaining the water insoluble compounds suspended as the sole carbon source for growth of the microorganism.

R. Parks, Brown University: There are two points that I would like to ask about. A year or so ago, there was a flurry of interest in observations that indicated that there is an increase in arylamine hydroxylase activity as one proceeds down the gastrointestinal tract. It is well known that tumors of the duodenum, although when they occur are highly malignant, are, in fact, very rare. We have heard reports of correlations between the activity and perhaps the inducibility of the P-450 cytochrome enzyme, an arylamine hydroxylase in the regions of the gastrointestinal tract most susceptible to malignancy. I wonder if you could bring us up to date on the present status of this very important question?

R. Estabrook: The effect of anti-oxidants on the system has been studied in great detail. I think people believe there is a correlation. By the way there is a rather high concentration of cytochrome P-450 in the intestinal mucosa that is readily induced. This has been overlooked by the pharmacologists.

R. Parks: Does the activity of the inducible enzyme increase the further down the GI tract that you do, in other words in correlation with the incidence with the bowel?

R. Estabrook: I honestly don't know.

R. Parks: The second question has to do with the current status of the observation reported in the New England Journal of Medicine a year or so ago, in which it was stated that a series of patients with bronchiogenic carcinoma had arylamine hydroxylase activity in their peripheral lymphocytes that were much more highly inducible than the enzyme in the lymphocytes of a comparable population of patients with diseases not related to pulmonary malignancy. Since the original report was published, I understand that these observations have been difficult to substantiate. I wonder whether you could comment on the present status of this very important problem.

R. Estabrook: I can only give you a comment using information I have received third hand. For those who did not understand the question, studies were carried out a few years ago which indicated that lymphocytes were capable of being induced by polycyclic hydrocarbons and separated into two classes; suggesting that there are two populations of individuals, i.e. one group with a readily inducible hydrocarbon hydroxylase and a second group less sensitive to induction. There was a meeting of the Tobacco Research Council where this was discussed in great detail. It appears that there were far too few samples in the study and it appears now that a normal distribution may be obtained. We will have to wait to learn further of the results of these studies.

W. Jakoby, National Institutes of Health: I would like to make a comment concerning another mechanism of in vivo detoxification which of course would not be present in your washed microsomes but was referred to by Dr. Weinhouse. That is the glutathione S-transferase group of enzymes which appear to be extremely important if only by reason of the enormous concentration of the protein themselves. In rat liver extracts, for example, they represent 10% of all soluble protein; in human liver they represent approximately 3% of all the soluble extractable protein. I would also like to comment on the fact that at least one of these proteins is inducible in liver and that another is inducible in gut. And in partial answer to Dr. Park's question, Larry Pincus, Jeanne Kethey and I have recently found that the higher up in the gut you are, the greater the concentration of glutathione S-transferase. The lower down you are, the less of the enzyme in correlation with the increased incidence of cancer in the lower gut.

R. Estabrook: I would certainly agree that glutathione can play a very critical role. In the scheme I presented, glutathione was indicated as one possible reactant for the epoxide. It is interesting to me that if you pretreat an animal with a carcinogen the liver very rarely develops tumors: the lung for example and other organs are much more susceptible. I don't know what the glutathione content is of such other organs.

W. Jakoby: It may not be the glutathione content, which is very high in all tissues, but rather the content of the transferases. As you know, they conjugate an enormous variety of compounds, including the 4-5 epoxide of benzpyrene.

MAGNETIC RESONANCE STUDIES OF THE MECHANISM OF DNA

POLYMERASE I FROM E. COLI

A.S. MILDVAN, D.L. SLOAN, C.F. SPRINGGATE AND L.A. LOEB
Institute for Cancer Research, Fox Chase
Philadelphia, Pennsylvania 19111

Abstract: DNA polymerases from animal, bacterial and viral sources contain Zn^{2+}, which has been established to be essential for catalysis in DNA polymerase I from E. coli and in the "reverse transcriptase" of avian myeloblastosis virus. Competition studies with o-phenanthroline and with Br^-, using nuclear quadrupolar relaxation, suggest that the Zn^{2+} functions in the binding of the DNA template-primer complex to the enzyme. DNA polymerases also require a divalent cation, such as Mg^{2+} or Mn^{2+} for activity. We have previously detected an active ternary $E-Mn^{2+}-$dTTP complex with E. coli DNA polymerase I by measurements of the paramagnetic effects of Mn^{2+} on the longitudinal relaxation rate of water protons. Using Mn^{2+} as a paramagnetic probe, we now find a unique conformation of dTTP in this ternary complex by ^{31}P and ^1H nuclear magnetic relaxation studies of dTTP. The calculated Mn^{2+}-to-phosphorus distances indicate that enzyme bound Mn^{2+} coordinates directly only with the γ phosphoryl group and that a puckered triphosphate conformation exists for the enzyme bound dTTP. This conformation differs from that of the binary $Mn^{2+}-$TTP complex in which α, β, and γ phosphoryl coordination occurs, and a thymine-deoxyribose torsion angle (χ) about the glycosidic bond of 40° is detected. The 8 Mn-substrate distances on the enzyme are fit by a unique Mn-dTTP conformation, with a torsion angle equal to 90°, indistinguishable from that found for a deoxynucleotidyl unit in double helical DNA-B. Hence binding to DNA polymerase appears to adjust the conformation of dTTP for Watson-Crick base pairing. When positioned within the double helical DNA structure, the conformation of enzyme-bound Mn-dTTP requires an in-line nucleophilic attack on the α-P. The role of Mn^{2+} appears to be facilitation of pyrophosphate departure, and possibly, activation of the α phosphoryl group. These mechanism studies provide clues to the design of inhibitors of DNA polymerase for possible use as chemotherapeutic agents in cancer.

INTRODUCTION

DNA polymerases catalyze the central reaction in the biosynthesis of DNA, the accurate copying of a polynucleotide template (1, 2).

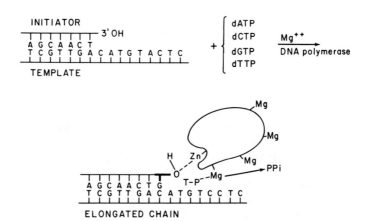

Fig. 1. Reaction catalyzed by DNA polymerase.

As indicated by Fig. 1, the reaction is a nucleophilic attack of the 3'-OH primer terminus upon the α phosphorus of deoxynucleoside triphosphate substrate with the displacement of pyrophosphate (1, 2). The reaction requires a template-primer complex, four deoxynucleotide substrates and a divalent cation activator such as Mg^{2+} or Mn^{2+}. In addition to the added divalent cation, we have found DNA polymerases from animal (3) bacterial (3, 4) and viral sources (4-6) to contain stoichiometric quantities of tightly bound Zn^{2+} (Table I). Our finding of Zn^{2+} in the DNA polymerase from avian myeloblastosis virus or "reverse transcriptase" (5, 6) has recently been confirmed (7) and extended to the polymerases of other RNA tumor viruses (8).

Of the various polymerase enzymes listed in Table I, in only two cases, E. coli DNA polymerase I (4) and AMV DNA polymerase (6) have we established the essentiality of Zn^{2+} for catalytic activity by demonstrating enzyme inactivation upon removal of Zn^{2+} and enzyme reactivation upon the replacement of Zn^{2+}. An essential role for Zn in all of the enzymes listed in Table I is suggested by the inhibition of

activity by the chelating agent, o-phenanthroline with little
or no inhibition by its non-chelating analog m-phenanthroline.
Hence it seems appropriate, at least in the case of E. coli
DNA polymerase I and AMV-DNA polymerase to refer to these as
Zn-metalloenzymes.

EXPERIMENTAL

Using kinetic and magnetic resonance methods described
in detail elsewhere (4, 11-14), we have been studying
the role of the bound Zn^{2+} and of the added Mn^{2+} in
catalysis by E. coli DNA polymerase I and other polymerases.
The results of these studies have provided useful clues to
the mechanism of action of DNA polymerases and to the
design of inhibitors of DNA polymerase for possible use
as chemotherapeutic agents in malignant disease.

RESULTS AND DISCUSSION

The Catalytic Role of Bound Zn^{2+} in DNA Polymerases.
Two indirect lines of evidence suggest that the bound Zn^{2+}
in DNA polymerase interacts with the DNA template-primer
complex. First, in most cases in which a kinetic analyses
of the inhibition by o-phenanthroline has been made, a
competitive component to this inhibition, (i.e., a slope
effect in a double reciprocal plot), with respect to
DNA, but not with respect to substrates has been detected.
These findings have been made with DNA polymerase I from
E. coli (3) and from sea urchin nuclei (3). However, with
AMV DNA polymerase, inhibition by o-phenanthroline is non-
competitive (6) which could be due either to inactivation
of the enzyme by rapid removal of the Zn^{2+} or to a
structural role of the Zn^{2+} in catalysis.

Second, the bound Zn^{2+} on E. coli DNA polymerase I has
a large effect on the nuclear quadrupolar relaxation rate
of $^{79}Br^-$ (4). This effect is diminished by 70% with
a sharp end point when one molecule of a polynucleotide
binds per enzyme molecule, but is unaltered by the substrate
dTTP suggesting that the polynucleotide, but not the
substrate, displaces 70% of the Br^- from the bound Zn^{2+} (4).

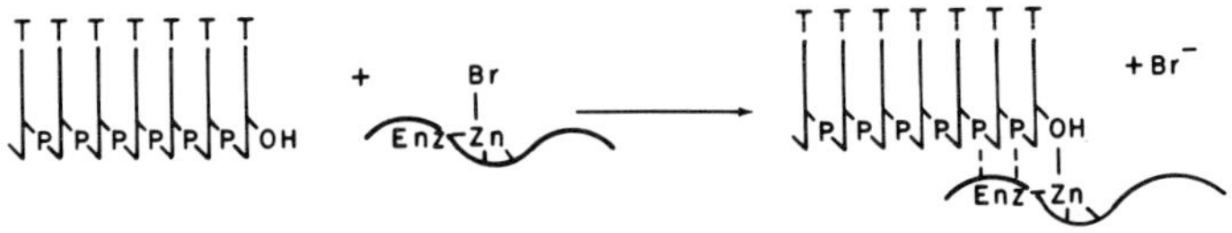

Fig. 2. Displacement of bromide ion from DNA polymerase-bound Zn^{2+} by the DNA analog $(dT)_{6-9}$ (4).

Because of their indirect competitive nature, these studies must be considered inconclusive but they do suggest a possible catalytic role for Zn^{2+}, namely coordination of the 3'-OH primer terminus (Fig. 2) facilitating its deprotonation and thereby preparing it to attack the α phosphorus atom of the substrate (4). This role for Zn^{2+} is analogous to that long considered for Zn^{2+} in carbonic anhydrase in the so-called Zn-hydroxide mechanism (9).

Regardless of the precise role of Zn^{2+}, its essentiality for DNA polymerase renders it a possible target for chemotherapeutic attack (10). The apparent universality of the zinc requirement for DNA synthesis argues against the ease of selectively interfering with DNA replication by animal tumor viruses or in malignant cells. However, certain chelators with a high affinity for Zn^{2+} could be selectively concentrated in particular types of malignant cells; alternatively, DNA replication in malignant cells with diminished concentrations of zinc may be more easily inhibited by chelating agents. Also, there may be quantitative differences in the rate of removal of zinc from viral and cellular polymerases. The results given in Fig. 3 suggest that DNA polymerases from RNA tumor viruses are more rapidly inactivated by ortho-phenanthroline than are cellular DNA polymerases (6). Thus it may be possible to selectively inhibit replication of RNA tumor viruses thereby interfering with the initiation and/or maintenance of malignant changes.

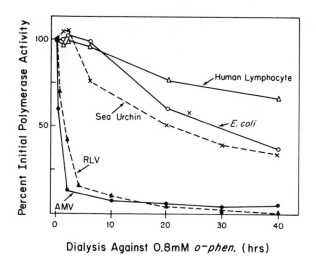

Fig. 3. Time-dependent inactivation of different DNA polymerases by 1,10 phenanthroline (6). (Note: RLV indicates Rauscher murine leukemia virus and AMV indicates avian myeloblastosis virus).

The Catalytic Role of the Added Divalent Cation in E. coli DNA Polymerase I. This problem has been investigated by replacing the diamagnetic Mg^{2+} with paramagnetic Mn^{2+} (11, 12). Binding studies using EPR reveal one tight binding active site for Mn^{2+} on the enzyme with a dissociation constant ($\sim$1 μM) in agreement with its kinetically determined activator constant (1.2 μM), four intermediate binding sites (K_D = 29 μM) and approximately 20 weak binding inhibitory Mn^{2+} sites with dissociation constants (0.8 mM) in agreement with the inhibitor constant of Mn^{2+} (0.6 mM) (11). Comparison of the binding and kinetic data indicates that occupancy of the tight site by Mn^{2+} activates the enzyme and occupancy of the weak sites inhibits activity. With Mg^{2+} as activator, a comparison of kinetic and binding properties indicates that the tight and intermediate sites must be occupied for enzyme activity, suggesting an essential role for the intermediate sites as well (11). In accord with this view nuclear relaxation studies of water indicate that deoxynucleotide substrates such as dTTP interact with Mn^{2+} at both the tight and intermediate sites to decrease the number of coordinated water ligands by $\sim$1 (11).

To examine in greater detail the structure and conformation of the ternary Pol I-Mn^{2+}-dTTP complex, the longitudinal and transverse nuclear relaxation rates of the phosphorus atoms and protons of the bound dTTP substrate were studied (12). The ^{31}P relaxation rates were measured at 40.5 MHz and those of the protons were measured at two frequencies, 100 MHz and 220 MHz, to permit evaluation of the correlation time. Varying concentrations of enzyme (24-100 µM), Mn^{2+} (0.2-60 µM); and of dTTP (5-10 mM) were used, permitting calculation of the relaxation rate of dTTP in the ternary complex (12).

From the theory of paramagnetic effects on nuclear relaxation rates (13, 14), the paramagnetic effect of a metal such as Mn^{2+} on the longitudinal relaxation rate ($1/T_{1p}$) of a magnetic nucleus such as ^{31}P or protons of a ligand in the same complex depends on four parameters of interest to a biochemist, the life-time of the complex, its stoichiometry, the correlation time for the electron-nuclear dipolar interaction, and the distance from the unpaired electron to the nucleus.

For the binary Mn-dTTP complex and the ternary DNA polymerase-Mn^{2+}-dTTP complex the transverse relaxation rates ($1/T_{2p}$) of the protons and ^{31}P of dTTP greatly exceeded the longitudinal relaxation rates ($1/T_{1p}$) indicating that the life-time contributes little to $1/T_{1p}$, thus eliminating one of the four unknowns. The correlation time for the binary complex (1.3 x 10^{-10} sec) was evaluated from $1/T_{1p}$ of water protons, and for the ternary complex (7 x 10^{-10} sec) from the frequency dependence of $1/T_{1p}$, eliminating the second unknown. The stoichiometry or coordination number of dTTP on Mn^{2+} in both the binary and ternary complexes was determined to be 1:1 from binding studies, eliminating the third unknown. The evaluation of these three parameters permitted calculation of the fourth, namely the distance from Mn^{2+} to five protons and 3 phosphorus atoms of dTTP in both the binary Mn-dTTP complex and the ternary DNA polymerase-Mn-dTTP complex (12). These distances were used to construct molecular stick models of both complexes (Fig. 4) (12). The uniqueness of these models was tested by constructing space filling models, and by a computer search among 47,000 conformations rejecting those structures which produced a total van der Waals overlap greater than 0.4 Å and which required distances which exceeded our error limits of 5.5% and 7.5% for the binary and ternary complexes, respectively. By these tests the structures of Fig. 4 provide a highly unique fit to our data (12).

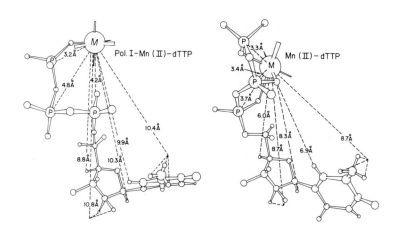

Fig. 4. Conformations and distances in the binary Mn^{2+}-dTTP complex and ternary Pol I-Mn^{2+}-dTTP complex as determined by nuclear relaxation (12).

Two major differences are noted between the binary and ternary complexes. First, in the binary complex all three of the phosphoryl groups of dTTP are directly coordinated to Mn^{2+}. On DNA polymerase only the γ-phosphoryl group remains coordinated by the enzyme-bound Mn^{2+}. The distance from Mn^{2+} to the β phosphorus atom (4.9 Å) indicates no direct coordination. The intermediate distance to the reaction center α phosphorus atom (4.2 Å) is most simply explained by the rapid averaging of $\leq 15\%$ inner sphere coordination with $\geq 85\%$ second sphere coordination, possibly with an intervening water ligand. The resulting polyphosphate conformation is puckered and somewhat strained. Hence an important role of the divalent cation activator in catalysis is to assist the departure of the leaving pyrophosphate group by γ-coordination, and possibly to facilitate nucleophilic attack on the α phosphorus atom by strain and by hydrogen bonding through a coordinated water ligand (12).

A second important difference between the binary and ternary complexes is in the conformation about the thymine-deoxyribose bond of dTTP. Such glycosidic conformations are quantitatively described by the torsion angle χ, in the present case the dihedral angle between N_1-C_6 of thymine and C_1'-O_1' of deoxyribose when viewed along the glycosidic N_1-C_1' bond. The χ value of $40 \pm 5°$ in the binary complex increases to $90 \pm 5°$ in the ternary complex (Fig. 4). Interestingly, the latter torsion angle of 90° is indistinguishable from that found for deoxynucleotidyl units in double helical DNA(15,16). Hence the binding of the substrate Mn-dTTP to the enzyme, DNA polymerase, in absence of template, has changed the substrate conformation to that of a nucleotidyl unit in the product-double helical DNA. Similarly, a 90° torsion angle is also found for the purine nucleotide substrate Mn-dATP when bound to DNA polymerase (12).

As discussed below this enzyme-induced change in substrate conformation may well represent an error preventing mechanism. When the structure of enzyme-bound Mn-dTTP (Fig. 4) is superimposed by computer onto the double-helical structure of DNA-B (Fig. 5), the resulting location of the α-phosphorus atom and the leaving pyrophosphate group of the bound substrate relative to the attacking 3'-OH group of the preceding nucleotide unit is consistent only with an in-line nucleophilic displacement on the α phosphorus (12). Hence the biosynthesis of nucleic acids, like their hydrolysis (17, 18) appears to proceed by an in-line mechanism.

The selection by the enzyme of those substrate conformations that fit into the double helix would amplify the Watson-Crick base pairing scheme and could thereby serve to prevent errors in template copying (12). Such error-preventing mechanisms are required by DNA polymerases (19) since enzymes from eukaryotic cells lack an error-correcting exonuclease activity (2) yet synthesize DNA with an accuracy at least two orders of magnitude greater than predicted by the thermodynamic (19) and kinetic effects (20) of base pairing alone. Error preventing mechanisms are either absent or less effective in error-prone DNA polymerases from AMV and other tumor viruses (21, 22) and from human leukemic lymphocytes (23). It has been suggested that the operation of such error prone DNA polymerases might be causally related to the progression of

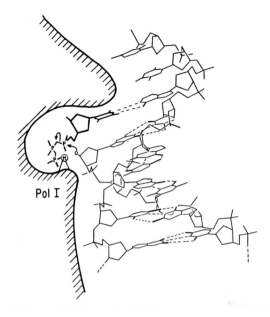

Fig. 5. Proposed mechanism of chain elongation catalyzed by DNA polymerase I consistent with the NMR data (12). The substrate, primer and template were positioned by superimposing by computer, the optimum conformation of Pol I-bound Mn^{2+}-dTTP illustrated in Fig. 4, onto that of a thymidylate residue of DNA. The role of the divalent cofactor, Mn^{2+} is shown. The role of the Zn^{2+} which is not shown is believed to be the binding and activation of the primer (4).

malignant disease (24). Possibly, these error prone enzymes fail to change the torsion angle of their substrates to 90° upon binding them. Rather, they may accept those substrate conformations which pre-exist free in solution such as the 40° conformation for Mn-dTTP. Thus a selective chemotherapeutic approach to malignant disease could involve the use of nucleotide inhibitors of DNA polymerase, the torsion angles of which are locked at 40°. These inhibitors might then show a preference for those error prone polymerases which may be associated with malignant disease.

Table 1

EVIDENCE THAT DNA POLYMERASES ARE ZINC METALLOENZYMES

Source	Zinc (gm-atoms/mole)	o-phenanthroline for 50% Inhibition (mM)	Ref.
Bacteria			
E. coli DNA Pol. I	1.8	0.04	(3)
E. coli DNA Pol. I	1.0 ± 0.15	-	(4)
Bacteriophage			
T_4 Phage	1.0	-	(4)
Eucaryotes			
Sea Urchin Nuclei	4.2	0.4	(3)
Human Lymphocytes	N.D.	1.0	(3)
RNA Tumor Viruses			
Avian Myeloblastosis	1.3	0.4	(5)
Avian Myeloblastosis	1.8 - 2.0	0.1	(7)
Murine Leukemia	1.4	0.012	(7)
Woolly Monkey	1.0	0.012	(7)
Related Enzymes			
Calf Thymus Terminal Transfer	N.D.	0.003	(25)
E.coli RNA Polymerase	2	0.04	(26)
Rat Liver RNA Polymerase	N.D.	0.025	(27)
T_7 RNA Polymerase	2 - 4	-	(28)

N.D. = not determined.

REFERENCES

1. T. Kornberg and A. Kornberg in The Enzymes (P.D. Boyer, Ed.), 3rd Ed., Vol. 5 (Academic Press, N.Y., 1971) p. 173.

2. L.A. Loeb in The Enzymes (P.D. Boyer, Ed.), 3rd Ed., Vol. 5 (Academic Press, N.Y., 1971) p. 119.

3. J.P. Slater, A.S. Mildvan and L.A. Loeb, Biochem. Biophys. Res. Commun. 44 (1971) 37.

4. C.F. Springgate, A.S. Mildvan, R. Abramson, J.L. Eagle and L.A. Loeb, J. Biol. Chem. 248 (1973) 5987.

5. B.J. Poiesz, N. Battula and L.A. Loeb, Biochem. Biophys. Res. Commun. 56 (1974) 959.

6. B.J. Poiesz, G. Seal and L.A. Loeb, Proc. Nat. Acad. Sci., USA 71 (1975) 4892.

7. D.S. Auld, H. Kawaguchi, D.M. Livingston and B.L. Vallee, Proc. Nat. Acad. Sci., USA 71 (1974) 2091.

8. D.S. Auld, H. Kawaguchi, D.M. Livingston and B.L. Vallee Biochem. Biophys. Res. Commun. 62 (1975) 296.

9. R.P. Davis in The Enzymes (P.D. Boyer, et al., Ed.) 2nd Ed., Vol. 5 (Academic Press, N.Y., 1961) p. 545.

10. J.W. Yarbro, C.H. Chang, A.S. Mildvan and L.A. Loeb Proc. American Assoc. for Cancer Research 13 (1972) Abs 65.

11. J.P. Slater, I. Tamir, L.A. Loeb and A.S. Mildvan J. Biol. Chem. 247 (1972) 6784.

12. D.L. Sloan, L.A. Loeb, A.S. Mildvan and R.J. Feldmann, J. Biol. Chem. 250 (1975) in press.

13. A.S. Mildvan and M. Cohn, Advances in Enzymology 33 (1970) 1.

14. A.S. Mildvan and J.L. Engle, Methods in Enzymology 26C (1972) 654.

15. S. Arnott and D.W.L. Hukins, Nature 224 (1969) 886.

16. M. Sundaralingam, Biopolymers 7 (1969) 821.

17. F.A. Cotton, E.E. Hazen in The Enzymes (P.D. Boyer, Ed.), 3rd Ed., Vol. 4 (Academic Press, N.Y., 1971) p. 153.

18. D.A. Usher, E.S. Erenrich and F. Eckstein, Proc. Nat. Acad. Sci., USA 48 (1972) 449.

19. A.S. Mildvan, Ann. Review of Biochemistry 43 (1974) 357.

20. E. Travaglini, A.S. Mildvan and L.A. Loeb, J. Biol. Chem. 275 (1975) 8647.

21. N. Battula and L.A. Loeb, J. Biol. Chem. 249 (1974) 4086.

22. M.A. Sirover and L.A. Loeb, Biochem. Biophys. Res. Commun. 61 (1974) 360.

23. C.F. Springgate and L.A. Loeb, Proc. Nat. Acad. Sci., USA 70 (1973) 245.

24. L.A. Loeb, C.F. Springgate and N. Battula, Cancer Res. 34 (1974) 2311.

25. L.M.S. Chang and F.J. Bollum, Proc. Nat. Acad. Sci., USA 65 (1970) 1041.

26. M.C. Scrutton, C.W. Wu and D.A. Goldthwait, Proc. Nat. Acad. Sci., USA 68 (1971) 2497.

27. P. Valenzuela, R.W. Morris, A. Faras, W. Levinson and W.J. Rutter, Biochem. Biophys. Res. Commu. 53 (1973) 1036.

28. J.E. Coleman, Biochem. Biophys. Res. Commun. 60 (1974) 641.

This work was supported by United States Public Health Service Grants AM-13351, RR-05539, CA-06927, CA-12818 (for use of NMR equipment located at the University of Pennsylvania), Grants BMS-74-03739 and BMS-73-06751 from the National Science Foundation, and by an appropriation from the Commonwealth of Pennsylvania.

Discussion

B. Vallee, Harvard University: I thought your comment on the time dependent inhibition of various "normal" polymerases as compared with those from "malignant" tissues might require some comment. Well, it turns out that Dr. Falchuk has just isolated the RNA polymerase II from E. gracilis which I shall discuss tomorrow and it happens to be inhibited instantaneously by 1,10 phenanthroline. Up to the present the only two enzymes I know for which this is true are the alcohol dehydrogenases from human and equine liver. So in that particular instance I would suspect certainly that the rate of inhibition of the polymerases of oncogenic viruses would be related to their effect on malignancy. I don't want to get ahead of what I have to say tomorrow, but it would seem from all the cell biological studies which we have recently performed by cell cytofluorometry, zinc seems to participate in cellular metabolism at all three stages, G1, S, G2 and in the correction of the mitotic apparatus and cell division. Thus it does not seem clear at the moment of what might be the limiting step in growth with respect to zinc. I presume in the end one will have to decide where and how agents like 1,10 phenanthroline effect these steps. That relates to your suggestion of chemotherapy. It is well taken but from a practical point of view the agents used for this purpose would have to be much more selective than 1,10 phenanthroline can be.

A. Mildvan, Institute for Cancer Research: Many chemotherapies result from lucky accidents. I am delighted that you agree that it is worth a try.

R. Estabrook, University of Texas: Can you tell us anything further concerning the binding of zinc in the enzyme? You implied that the zinc could be removed and reinserted. Is it possible to use other metal ions, such as the rare earth elements, to replace the zinc? This might then give you the opportunity to measure any interaction between paramagnetic ions as they reside on the enzyme. Second, it was unclear to me how you ruled out dinucleotide formation in which the divalent cation served as a linking bridge.

A. Mildvan: O.K. To answer your first question it is unusual for a rare earth to activate an enzyme and only then can one be sure that its properly binding. Amylase is one of the few examples. We have shown that replacement of zinc by cobalt gives about half maximal activity so this might be a more useful paramagnetic probe of the zinc site on DNA polymerase I. Now your second question. This is one of the reasons we did a study using variable substrate concentration. At the nucleotide concentrations we used on the order of 5mM such problems are not observed, nor were any such products detected by NMR.

S. Greer, University of Miami: Regarding the low IQ of the tumor virus polymerases, well this is interesting. It is completely consistent with Temin's idea on the misevolution hypothesis regarding the origin of the oncogene. Could you comment on this?

A. Mildvan: Well it is difficult with this enyzme only because we have so little of it. But they both interact with the same tightly bound manganese. In the same geometric way, so by those criteria they are at the same place. We could test that by determining whether one nucleotide could displace the other and that experiment has not yet been done by NMR.

R. Parks, Brown University: I assume that when you refer to the potential chemotherapeutic implications of your observations, you have considered the possible application to combination chemotherapy. Your finding of apparently marked differences in the DNA polymerases from normal tissue as compared with cancer tissues, of course, raises the hope that one might be able to design specific methods of inhibiting the cancer-specific enzyme, but not the normal tissue enzyme. If such a feat could be accomplished, it could prove of greatest value if used in combination with other modalities that cause specific damage to the DNA molecule such as irradiation therapy or the alkylating agents. One might expect to observe a marked synergistic effect if a DNA damaging agent could be combined with the specific inhibition of major component of the DNA repair enzyme systems of the malignant, but <u>not</u> of the normal tissues.

A. Mildvan: I am no expert on chemotherapy at all but I would simply say that this might serve some supplemental role to existing therapeutic agents of all types.

R. Block, Papanicolaou Cancer Research Institute: In recent years, Dr. Bothner-By has investigated the use of the nuclear Overhauser effect to study binding sites in cases of small molecules binding to polypeptides. I was wondering if you think this kind of approach might be used to study the interaction of phenanthroline at the zinc site?

A. Mildvan: Very briefly let me respond to that by saying that the nuclear Overhauser effect depends upon small changes in the energy levels of the magnetic nuclei involved. This means that what one would observe would be small effect at best. It's worth a try but you are going to have to push signal-to-noise ratio to a great extent to do this type of experiment.

ENZYMES INVOLVED IN REPAIR OF DNA DAMAGED BY CHEMICAL CARCINOGENS AND γ-IRRADIATION[1]

DOLLIE M. KIRTIKAR, J. PHILIP KUEBLER,
ANTHONY DIPPLE[2] AND DAVID A. GOLDTHWAIT[3]
Department of Biochemistry
Case Western Reserve University
Cleveland, Ohio 44106
and
The Chester Beatty Research Institute
Fulham Road, London, U.K.

Abstract: Endonuclease II of E. coli, active on DNA treated with methyl methanesulfonate (MMS), methylnitrosourea (MNU), 7-bromomethyl-12-methylbenz[a]anthracene (7BMMB), and γ-irradiation, has been separated from an endonuclease active on depurinated DNA. The mutant AB3027 lacks endonuclease II while the mutant BW2001 lacks the apurinic acid endonuclease. Endonuclease II has a molecular weight of approximately 33,000, has no divalent metal requirement but is stimulated by Mg^{++}. The products of phosphodiester bond hydrolysis in MMS DNA are 3' hydroxyls and 5' phosphates. Endonuclease II hydrolyzes phosphodiester bonds and also releases 3-methyladenine, but not 7-methylguanine from DNA treated with dimethylsulfate. From MNU treated DNA, the enzyme releases O-6-methylguanine and 3-methyladenine as well as 1-methyl- and 7-methyladenine. DNA treated with 7-bromomethyl-12-methylbenz[a]anthracene is a substrate for endonuclease II. The enzyme hydrolyzes phosphodiester bonds and releases the N^6 adenyl and N^2 guanyl derivatives. Phorbol ester inhibits both phosphodiester bond hydrolysis and base release from 7BMMB treated DNA catalyzed by endonuclease II. Phorbol ester does not inhibit the apurinic acid endonuclease. Endonuclease II recognizes damage in DNA due to γ-irradiation. Enzyme sensitive sites can be increased in the DNA by a preincubation at 37° for 4 hours in the presence of nitrogen after the irradiation. An endonuclease specific for apurinic acid has been purified from calf liver. A comparison of some of the properties of this enzyme with the enzyme from calf thymus suggests that they are isoenzymes.

Endonuclease II of E. coli has, because of recent work, been defined as the enzyme activity which recognizes DNA which has been reacted with methylmethane sulfonate (MMS)(1), methylnitrosourea (MNU) (2), 7-bromomethyl-12-methylbenz[a]-anthracene (3) and γ-irradiation (4). This enzyme has now been separated from an enzyme active on depurinated DNA which will be referred to as the apurinic acid endonuclease. This latter enzyme has been studied by Verly and his collaborators who have purified it to homogeneity (5).

ENDONUCLEASE II AND THE APURINIC ACID ENDONUCLEASE OF
E. COLI-SEPARATION AND CONTROL BY GENES

Endonuclease II was separated from the apurinic acid endonuclease by chromatography on DEAE. The E. coli cells were disrupted with glass beads, and after a high speed centrifugation, treatment with 0.8% streptomycin and ammonium sulfate precipitation between 45% and 80% (6), the fraction was applied to a DEAE column and eluted as shown in Fig. 1.

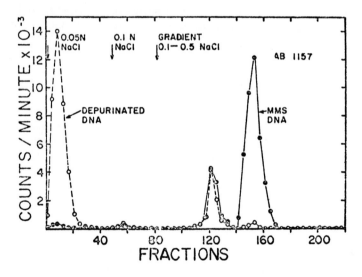

Fig. 1

Peak I is active on depurinated DNA or depurinated-reduced DNA (7) and is the same enzyme as that isolated by Verly and Paquette (8). This peak, now purified over 3000 fold has very slight activity on MMS treated DNA which may be due to minimal depurination, although this is difficult to prove. The molecular weight of this fraction after further purification was 33,000 as determined by gel filtration.

This corresponds well with molecular weights obtained by Verly and Rassart (5) of 33,000 by gel filtration and 32,500 by SDS gel electrophoresis. Peak III is active on DNA treated with MMS but is able to degrade depurinated DNA at a rate which is less than 10% of the rate of degradation of MMS treated DNA. This activity on depurinated DNA compared to MMS DNA remained constant through various steps of purification and evidence will be presented in another section to support the concept that this is an intrinsic activity of endonuclease II and not a contamination with the apurinic acid enzyme. In previous publications it was concluded erroneously that the activities on depurinated DNA and on MMS DNA were due to the same enzyme. Peak II contains both activities and on gel filtration has a molecular weight of 58,000. On rechromatography the same peak is recovered and does not split into Peaks I and III. This suggests that it is a stable dimer.

Using the chromatographic procedure shown above, we have examined two mutants and found that these mutations correspond to Peaks I and III. A mutant of E. coli originally isolated by Dr. Paul Howard-Flanders, AB 3027, on the basis of its sensitivity to MMS, has been shown to lack endonuclease II, Peak III, as illustrated in Fig. 2. Of note

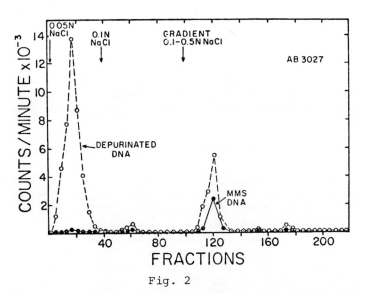

Fig. 2

is the activity in Peak II on MMS-DNA which is decreased to approximately one half that on the depurinated DNA. A second mutant, BW 2001, was isolated in Dr. Bernard Weiss' laboratory on the basis of a decreased endonucleolytic activity on DNA treated with MMS (9). The ammonium sulfate fraction of this mutant when examined on a DEAE column produced the pattern shown in Fig. 3. Peak I, the apurinic

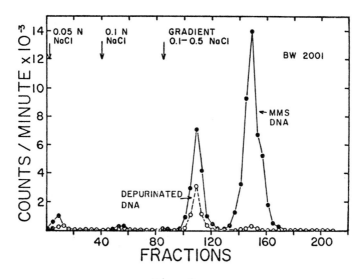

Fig. 3

acid activity, is absent and this activity is decreased to approximately one half in Peak II. Our interpretation of Peak II is that monomers of either enzyme can combine to form, for reasons unknown, a stable dimer. If one polypeptide has a missense mutation it is unstable as a monomer but is partially stabilized by the normal polypeptide in the dimer.

ENDONUCLEASE II PROPERTIES

The molecular weight by gel filtration of the material in Peak III is 33,000. When MMS treated DNA is used as the substrate, there is no requirement for divalent cation, although the activity is increased approximately two fold by Mg^{++} (1). The enzyme is active in the presence of various chelating agents such as 8-hydroxyquinoline of 6×10^{-4} M, but EDTA does inhibit the enzyme 71% at 10^{-4} M and 90% at 10^{-3} M. t-RNA does not inhibit the enzyme. The sulfhydryl reagent, p-choromercurisulfonate, does inhibit the enzyme activity. By examining the DNA in neutral and alkaline gradients it has been noted that the enzyme makes one double strand break for every four single strand breaks. Since the enzyme had very limited activity on untreated single-stranded DNA, these results suggest that the DNA may be alkylated in a non-random fashion.

The endonuclease II (Peak III) has now been purified over 3000 fold and with this preparation the nature of the phosphodiester bond break has been studied using snake venom and bovine spleen exonucleases, with and without alkaline phosphatase. Results are presented in Fig. 4 and from this

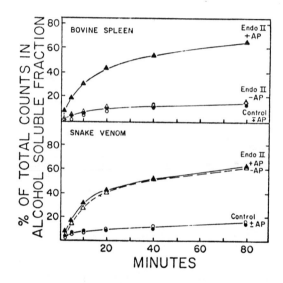

Fig. 4

it can be concluded that the enzyme produces 3'-hydroxyl and 5'-phosphate residues. Similar experiments have been done with the apurinic acid endonuclease and similar results have been obtained.

Because of the report that endonuclease II and exonuclease III were the same enzyme (9), the endonuclease II was examined for exonuclease III using as an assay the liberation of 3'-phosphate groups labeled with ^{32}P. On Sephadex G-100, a clear cut separation of the main endonuclease and exonuclease activities was found. The molecular weight of the exonuclease was approximately 26,000. No data has been obtained on the apurinic acid endonuclease, and its relationship to exonuclease III.

N-GLYCOSIDASE ACTIVITY OF ENDONUCLEASE II--THE RELEASE OF METHYLATED PURINES FROM DNA TREATED WITH DIMETHYLSULFATE OR METHYLNITROSOUREA

Endonuclease II preparations not only hydrolyze phosphodiester bonds, but also have an N-glycosidase activity. This is not a general activity, but to date is limited to purine derivatives and only to certain ones. When DNA was treated with [^{3}H-methyl] dimethylsulfate and then reacted with an endonuclease II preparation, 3-methyladenine but not 7-methylguanine was liberated and identified by paper and column chromatography (2). The ratio of 3-methyladenine to 7-methylguanine in the treated DNA was approximately 1:4. Thus the enzyme shows some specificity.

DNA was then reacted with methylnitrosourea (MNU) which is a strong mutagen and carcinogen. This agent produces not only 3-methyladenine and 7-methylguanine, but also O-6 methylguanine and small amounts of 1-methyl- and 7-methyladenine (10). O-6 methylguanine is now considered to be the altered base responsible for mutations (11) and incorrect base pairing has been demonstrated in vitro (12). A preparation of endonuclease II recognizes altered bases in MNU treated DNA and makes phosphodiester bond breaks and liberates free bases (2). The enzyme releases O-6 methylguanine, 3-methyladenine, and small amounts of 1- and 7-methyladenine. Some of the bases can be isolated by chromatography on Dowex 50 and a balance study is shown in Table I. The base re-

Table I

Stoichiometry of the enzyme reaction with DNA alkylated with MNU

	Percentage of the total counts in the reaction mixture					
	Alcohol-soluble fraction			Alcohol-insoluble fraction		
Fractions	− Enzyme	+ Enzyme	Δ	− Enzyme	+ Enzyme	Δ
O^6-methylguanine	1.4	6.8	+5.4	7.8	2.8	−5.0
3-methyladenine	1.8	13.7	+11.9	21.1	7.0	−14.1
7-methylguanine plus 7-methyladenine	7.0	9.0	+2.0	41.8	37.8	−4.0

−Enzyme, without endonuclease II in the reaction mixture.
+Enzyme, with endonuclease II in the reaction mixture.
Details of this work have been published (2).

leased in the peak which contains both 7-methyladenine and 7-methylguanine is 7-methyladenine. Thin layer chromatography was also used to examine the release of bases as a function of enzyme concentration. Results are shown in Table II. Again it is clear that there is no enzymatic re-

Table II

The Percent of the Total Counts Present on a Thin-Layer Chromatogram Recovered in Different Bases as a Function of Enzyme Concentration

	Enzyme Units			
	0.02U % total cpm	0.04U %	0.08U %	0.20U %
3-methyladenine	5.0	10.7	14.2	14.8
O^6-methylguanine	0.5	2.3	3.0	5.5
1-methyladenine	0.41	0.89	1.2	1.4
7-methyladenine	0	0.1	0.7	1.2
7-methylguanine	0.03	0	0	0

Unlabeled T4 DNA (3000 nmol) in 2.0 ml was reacted with [^{3}H]MNU (specific activity 48 mCi/mM) at MNU:DNA nucleotide ratio of 7:1; after alcohol precipitation and washing the alkylated DNA was solubilized in 1.0 ml of 0.05 M Tris·HCl (pH 8.0) at 0°. For enzymatic hydrolysis, 0.1 ml reaction mixtures containing 20 nmol [^{3}H]-MNU-T4 DNA (specific activity 2500 ^{3}H cpm/nmol of nucleotide, with 23.7 nmol of [^{3}H]methyl per mol of DNA nucleotide), 1 x 10^{-4} M β-mercaptoethanol, 1 x 10^{-4} M 8 hydroxyquinoline, 5 x 10^{-2} M Tris·HCl (pH 8.0), and enzyme as indicated were incubated at 37° for 1 hr. The reactions were terminated with EDTA at a final concentration of 0.02 M. The aliquot of 20 μl (approximately 5000 ^{3}H cpm), supplemented with methylated bases, from each sample was chromatographed separately on thin-layer sheets. The total counts recovered from all the sections of the thin-layer plate were taken as 100%. Details of this work have been published (2).

lease of 7-methylguanine. When the base release was examined as a function of time, it was found that 3-methyladenine was released at a rate 3-4 times that of the other bases. These in vitro experiments check very well with the in vivo results of Lawley and Orr (13) who found in MNU treated cells, loss of O-6 methylguanine and 3-methyladenine from the DNA, but no loss of 7-methylguanine. It is reasonable to assume that endonuclease II was responsible for the in vivo results. These in vitro results show that the enzyme has a phosphodiester bond hydrolytic activity as well as an N-glycosidase activity which has some specificity for purine derivatives. The stoichiometry between base release and phosphodiester bond cleavage has not been examined.

7-BROMOMETHYL-12-METHYLBENZ[A]ANTHRACENE TREATED DNA, A SUBSTRATE FOR ENDONUCLEASE II

7-Bromomethyl-12-methylbenz[a]anthracene was synthesized originally by Dipple and his colleagues and found to be a potent carcinogen in several animal test systems (14). It reacts with the amino groups of adenine, guanine and cytosine of DNA both in vitro and in vivo (15, 16). We have shown that the enzyme endonuclease II makes phosphodiester bond breaks and also releases the hydrocarbon-purine base derivatives of DNA treated with the brominated hydrocarbon (3). The relationship of phosphodiester bond breaks to enzyme concentration is shown in Table III. The enzyme

Table III

Enzyme induced single strand breaks in DNA treated with 7-bromomethyl-12-methylbenz[a]anthracene.

Enzyme Units	Single Strand Breaks
--	3
0.01	3
0.02	6
0.04	11
0.08	16
0.16	25

[^{3}H] purine-labeled T7 DNA [sp. act. 1960 cpm per nmol DNA nucleotide] was reacted with unlabeled 7-bromomethyl-12-methylbenz[a]anthracene at a hydrocarbon to DNA nucleotide ration of 1:10 (13). Incubation mixtures (0.25 ml) contained 15 nmoles of hydrocarbon modified DNA nucleotides, 1 x 10^{-4} M β-mercaptoethanol, 1 x 10^{-4} M 8-hydroxyquinolene, 5 x 10^{-2} M Tris·HCl buffer, pH 8.0 and enzyme units as indicated. After 60 minutes at 37°, reactions were terminated by adding EDTA and sodium dodecylsulfate at final concentrations of 2 x 10^{-2} M and 0.25% respectively. The samples were then incubated in alkali [0.066 M final concentration] at 37° for 20 minutes and aliquots were centrifuged through 5-20% alkaline sucrose density gradient solutions. Single strand breaks were calculated.

catalyzed release of the purine base derivatives is shown in Fig. 5 and in Table IV. After an incubation period with or without enzymes, alcohol was added to the reaction mixture to precipitate the DNA. The supernatant or alcohol soluble fraction which contained bases liberated by the enzyme was then chromatographed on Sephadex LH-20. The alcohol precipitate was hydrolyzed enzymatically to yield

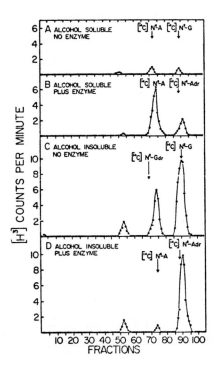

Fig. 5

nucleosides and then hydrolyzed with acid to obtain the base derivatives which were also chromatographed. The chromatographic patterns of one experiment are shown in Fig. 5. By the use of marker derivatives and by subsequent examination of the peaks by thin layer chromatography it was demonstrated that the derivatives of adenine and guanine, the free bases and not the nucleosides, are liberated. Good stoichiometry of release of the purine derivatives is shown in Table IV. Further experiments showed that the rate of release of the adenine derivative was approximately four times that of the guanine derivative. From these experiments it is concluded that endonuclease II can recognize DNA treated with 7-bromomethyl-12-methylbenz[a]anthracene and break phosphodiester bonds as well as liberate N^6 [12-methylbenz[a]anthracenyl-7-methyl] adenine and N^2 [12-methylbenz[a]anthracenyl-7-methyl] guanine. No evidence for liberation of the cytosine derivative was obtained, and this suggests that the enzyme may be specific for purine derivatives.

Table IV

ENZYMATIC RELEASE OF DMBA DERIVATIVES OF DNA BASES

Base Derivative	Alcohol Soluble Δ(+)vs(-) Enzyme CPM	Alcohol Insoluble Δ(+)vs(-) Enzyme CPM
N-6 DMBA Adenine	+22,423	-21,590
N-2 DMBA Guanine	+ 6,764	- 6,757
N-4 DMBA Cytosine	- 89	- 512

The incubation mixtures (1.0 ml) contained 85 nmoles DNA nucleotide [(^{3}H)-hydrocarbon-salmon sperm DNA, 0.82 mmoles carcinogen permole DNA-nucleotide, specific activity 500 cpm per nmole DNA nucleotide], 1 x 10^{-4} M β-nercaptoethanol, 5 x 10^{-2} M Tris HCl, pH 8 buffer, 1 x 10^{-4} M 8-hydroxyguinoline and 0.4 units of enzyme where indicated. After 60 minutes at 35° the reaction was terminated, and an alcohol soluble and an alcohol insoluble fraction were obtained. The latter was digested to nucleosides (15) and hydrolyzed to bases. The hydrolyzed sample as well as the alcohol soluble material were supplemented with [^{14}C]-labeled markers and chromatographed on Sephadex LH-20 eluted in spectro-analyzed methanol.

INHIBITION OF ENDONUCLEASE II BUT NOT THE APURINIC ACID ENDONUCLEASE BY PHORBOL ESTER

Phorbol myristate acetate, the active cocarcinogen in croton oil (17) inhibits both the hydrolysis of phosphodiester bonds and the N-glycosidase activity of endonuclease II. As shown in Fig. 6, the phorbol ester inhibits phospho-

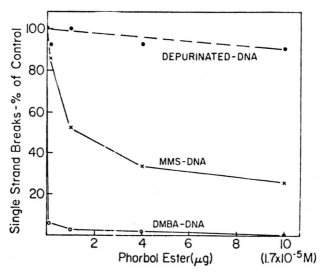

Fig. 6

diester bond hydrolysis of DNA treated with MMS or with 7-bromomethyl-12-methylbenz[a]anthracene. The hydrolysis of phosphodiester bonds in depurinated DNA by the apurinic acid endonuclease is not significantly inhibited by phorbol ester. The preparation used for these experiments contained both enzyme activities. The marked sensitivity of the hydrocarbon treated DNA (with 95% inhibition at 3.3×10^{-7} M) compared to the less sensitive and incomplete inhibition of the hydrolysis of MMS treated DNA is of note. The lack of complete inhibition could be due to some depurination, but the fact that higher concentrations of the phorbol ester were required for 50% of the maximal inhibition suggests that the K_m for the two substrates may be different. Release of hydrocarbon-labeled bases from DNA reacted with 7-bromomethyl-12-methylbenz[a]anthracene was also inhibited by phorbol myristate acetate. The percent inhibition of release of the bases as a function of concentration of the ester is shown in Table V. Again the sensitivity to very low concentration is apparent. Careful studies at these low levels comparing inhibition of phosphodiester bond hydroly-

Table V

INHIBITION OF ENZYMATIC RELEASE OF
DMBA-BASE BY PHORBOL ESTER

Phorbol Ester µg	% Inhibition
0	0
0.02	51
0.10	92
1.0	96
5.0	99
10.0	99

(1 µg = 1.66×10^{-6} M)

Incubation mixtures (0.25 ml) contained 60-75 nmoles [^{3}H] hydrocarbon-treated DNA, 5×10^{-2} M Tris-HCl, pH 8.0 buffer, 1×10^{-4} M β-mercaptoethanol, 1×10^{-4} M 8 hydroxyquinoline, phorbol ester as indicated. Enzyme (0.06 units for 0.02 and 0.1 µg of phorbol ester and 0.25 units for 1-10 µg) was reacted with phorbol ester 5 minutes at 0° prior to addition of substrate. After 45 and 60 minute incubation at 36° with the two enzyme concentrations, reactions were terminated by adding EDTA to final concentration of 2×10^{-2} M. The samples were supplemented with unlabeled T4 DNA (50-100 µg per tube) and unhydrolyzed DNA was precipitated out with alcohol. Alcohol soluble and alcohol insoluble radioactivity was determined.

sis with inhibition of base release have not yet been done. Also the experiments described here have been done with a mixture of endonuclease II and the apurinic acid endonuclease although other experiments have shown that all the base release can be ascribed to endonuclease II and none to the apurinic acid endonuclease.

THE MECHANISM OF ACTION OF ENDONUCLEASE II

The problem still remains whether endonuclease II is a single enzyme with both the N-glycosidase and phosphodiester hydrolase activities. On purification these activities remain together, but the preparation is not yet homogeneous. Three other lines of evidence make it unlikely that there is a separate N-glycosidase which acts in conjunction with an enzyme similar or the same as the apurinic acid endonuclease. First is the evidence obtained with mutant AB 3027. If the mutation involved only the loss of an N-glycosidase then there should be some apurinic acid endonuclease present. One would also predict that if this was a complex the molecular weight should be greater than that of the apurinic acid endonuclease. Second, if the reaction was due to two separate enzymes then one of these enzymes should be rate limiting for the overall reaction. Careful studies with defined substrates--MMS and depurinated DNA's--have shown that the rate with depurinated DNA is always much slower than with MMS DNA. This could occur if endonuclease II has two sites--one an N-glycosidase and the second a phosphodiester bond hydrolase--and MMS DNA has a lower K_m than the apurinic acid. Third, support for a single protein comes from studies with phorbol ester. There is no inhibition by phorbol ester of hydrolysis of apurinic acid by the apurinic acid endonuclease. There is, however, inhibition by phorbol ester of the low level of hydrolysis of apurinic acid by endonuclease II. This might be expected if phorbol ester binds to a site on the enzyme, which recognizes purine derivatives and this bulky compound inhibits the binding of apurinic acid to the phosphodiester bond hydrolysis site. It is evident that further purification and studies are required to clarify the mechanism of base release and phosphodiester bond hydrolysis.

ACTION OF ENDONUCLEASE II ON γ-IRRADIATED DNA

Damage in DNA due to γ-irradiation is also recognized by endonuclease II (4). It has been shown that this activity is associated with the protein in Peak III. Two types of radiation induced single strand breaks have been measured in the experiments described here. The first type is due to an enzymatic hydrolysis at a specific site(s): the second type is due to a chemical hydrolysis of an alkali sensitive bond. The experiments are complicated by a comparison of the effects of O_2 and N_2 and by the finding that under certain conditions both types of sites would increase during an incubation following the irradiation.

The DNA was irradiated under N_2 (Fig. 7A, 7B) or O_2 (Fig. 7C) with 22.5 Krads from a ^{60}Co source. The DNA was then preincubated in O_2 (Fig. 7A, 7C) or N_2 (Fig. 7B) for up

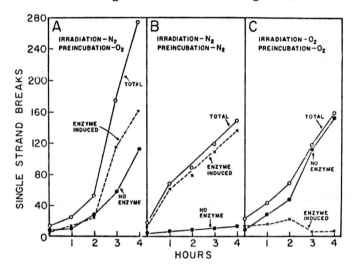

Fig. 7

to 4 hours. Samples were removed at varying times and then incubated with or without enzyme for one hour, exposed to 0.066 N NaOH at 37° for 20 minutes to hydrolyze any alkali sensitive sites and then examined in alkaline sucrose gradients. The alkali treatment would convert any depurinated or depyrimidinated sites to phosphodiester bond breaks which would be registered as spontaneous and not enzyme induced breaks. The DNA irradiated under N_2 and then preincubated with O_2 (Fig. 7A) showed an increase in both enzyme sensitive sites and in spontaneous breaks. The DNA irradiated under N_2 and then preincubated in N_2 showed a linear increase in enzyme sensitive sites, but no increase in spontaneous breaks. This indicates that spontaneous breaks are dependent upon O_2, but in these experiments as opposed to those of others, the O_2 effect has been separated from the initial effects of irradiation. This suggests that at least two types of irradiation damage occurs. The first type produces some form of metastable structure which on incubation goes on to become an enzyme sensitive site. This is not a depurinated or depyrimidinated site. This is recognized by endonuclease II and because of the other substrates recognized by this enzyme, we would propose that the damage involves a purine base or bases. This is currently under investigation.

The second type of damage is also due to some metastable product possibly a DNA-hydroxyl radical which can either decay to a stable product in N_2 or which, because of its long half life, can react with O_2 to form a more unstable product possibly an altered base. This product could be released to give a depurinated or depyrimidinated site. Evidence to support this scheme has been obtained by the use of $NaBH_4$ and NH_2OH as shown in Table VI. Previous work had shown that either of these reagents could be used

Table VI

The Effect of Addition of $NaBH_4$ or NH_2OH before or after Preincubation of Irradiated DNA on Spontaneous Single Strand Breaks.

Preincubation	Additions Before Preincubation	After Preincubation	Single Strand Breaks
+	---	---	47
+	$NaBH_4$	---	4
+	---	$NaBH_4$	11
+	NH_2OH	---	5
+	---	NH_2OH	13
-	---	---	8
-	---	$NaBH_4$	4
-	---	NH_2OH	4

Vials containing [^{3}H] thymine labeled T7 DNA (2900 cpm/nmol of DNA nucleotide) was irradiated under nitrogen at pH 7.0, 0° C, by a ^{60}Co γ-ray source with 22.5 Krads. The pH was adjusted to 8.5 and some samples were preincubated under nitrogen at 37° for four hours. Additions of $NaBH_4$ at a final concentration of 0.25 M or NH_2OH at a final concentration of 0.2 M were as indicated. Prior to analysis by sucrose gradient centrifugation all samples were exposed to 0.066 N NaOH at 37° for 20 minutes conditions adequate to hydrolyze phosphodiester bonds at depurinated sites. Details of the conditions have been reported (4).

to stabilize depurinated sites to alkali (7). After irradiation under N_2, either of these agents could be added at the beginning or at the end of a preincubation at 37° in O_2 and they would prevent most of the increase of the alkali labile bonds. Although we have demonstrated that label from both purine and pyrimidine labeled DNA is released during the incubation we do not have definitive evidence that base loss is responsible. An oxidation reaction on the sugar could produce the same results. Experiments with the apurinic acid endonuclease may help to define the nature of these alkali labile bonds.

AN APURINIC ACID ENDONUCLEASE FROM CALF LIVER

From calf liver an endonuclease specific for depurinated DNA has been purified approximately 900 fold. The enzyme does not recognize the sites in γ-irradiated DNA which are recognized by endonuclease II, nor does the enzyme recognize thymine dimers. Although one might predict that it was the same enzyme as that isolated from calf thymus by Ljungquist and Lindahl (18), an examination of the properties of this enzyme showed that it differed in many respects from the thymus enzyme. In Table VII a comparison of some of the properties of this enzyme with the properties of the thymus enzyme purified 830 fold by Ljungquist and Lindahl (19) and 10-30 fold in this laboratory. The pH optimum of the calf liver enzyme is higher than the calf thymus

Table VII

Calf Apurinic Acid Enzymes

Properties	Liver (Kuebler)	Thymus (Lindahl)	Thymus (Kuebler)
Purification	900	830	10-30
pH optimum	9.5	8.5	----
Max. Mg^{++} stimulation	1.3 x	1,000X	8X
Conc. Mg^{++} for max. stimulation	0.01 mM	3.0 mM	0.5-2.0 mM
50% inhibition by salt	0.023 M	0.2 M	0.12 M
Stimulation by low salt	none	---	> 4 fold at 0.01 M
DNA breaks	SS and ds	SS	----
M.W.	29,000	32,000	

enzyme and it does not show the marked dependence on divalent metal shown by the thymus enzyme. The liver enzyme, although it shows some stimulation by Mg^{++} at 0.01 M, is inhibited by higher concentrations which are required for maximum stimulation of the thymus enzyme. Unlike the thymus enzyme, the liver enzyme is very sensitive to ionic strength, and is inhibited 50% by 0.023 M NaCl. The thymus enzyme is stimulated by low concentrations of NaCl and inhibited 50% only by 0.12-0.2 M NaCl. The liver enzyme makes both single and double strand breaks in depurinated reduced DNA while the thymus enzyme is reported to make only single strand breaks. The molecular weights of these enzymes are not exactly the same. The liver enzyme has a molecular weight of 29,000 by gel filtration while the thymus enzyme is reported to be 32,000. The differences between these properties of the calf liver and calf thymus apurinic acid endonucleases are enough to support the concept that these enzymes are isoenzymes.

FOOTNOTES

[1]Supported by grants from the National Institutes of Health (CA11322), the Health Fund of Greater Cleveland, the Cuyahoga Unit of the American Cancer Society, and a contract with ERDA (11-1) 2725.
[2]Present address, Frederick Cancer Research Center, Frederick, Maryland 21701.
[3]National Institutes of Health Research Career Award Fellowship K6-GM-21444.

REFERENCES

(1) E.C.Friedberg, S.M.Hadi, and D.A.Goldthwait. J. Biol. Chem. 244 (1969) 5879.
(2) D.M.Kirtikar and D.A.Goldthwait. Proc. Nat. Acad. Sci. 71 (1974) 2022.
(3) D.Kirtikar, A.Dipple, and D.A.Goldthwait. Biochemistry (in press).
(4) D.Kirtikar, J.Slaughter, and D.A.Goldthwait. Biochemistry 14 (1975) 1235.
(5) W.G.Verly and E.Rassart. J. Biol. Chem. (1975) (in press).
(6) E.C.Friedberg and D.A.Goldthwait. Proc. Nat. Acad. Sci. 62 (1969) 934.
(7) S.M.Hadi and D.A.Goldthwait. Biochemistry 10 (1971) 4986.
(8) W.S.Verly and Y.Paquette. Canad. J. Biochem. 50 (1972) 217.
(9) D.M.Yajko and B.Weiss. Proc. Nat. Acad. Sci. 72 (1975) 688.
(10) P.D.Lawley and S.A.Shah. Biochem. J. 128 (1972) 117.
(11) A.Loveless. Nature 223 (1969) 206.
(12) L.L.Gerchman and D.B.Ludlum. Biochim. Biophys. Acta 308 (1973) 310.
(13) P.D.Lawley and D.J.Orr. Chem. Biol. Interactions 2 (1973) 154.
(14) F.J.C.Roe, A.Dipple, and B.C.V.Mitchley. Br. J. Cancer 26 (1972) 461.
(15) M.P.Rayman and A.Dipple. Biochemistry 12 (1973) 1202.
(16) M.P.Rayman and A.Dipple. Biochemistry 12 (1973) 1538.

(17) E.Hecker. Methods Cancer Res. 6(1971)439.
(18) S.Ljungquist and T.Lindahl. J. Biol. Chem. 249(1974) 1530.
(19) S.Ljungquist, A.Anderson, and T.Lindahl. J. Biol. Chem. 249(1974)1536.
(20) A.Dipple, P.Brookes, D.S.Mackintosh, and M.P.Rayman. Biochemistry 10(1971)4323.

Discussion

J.Van Lancker, University of California, L.A.:
Bacterial models have often proved useful to studying cancer, and this has surely been the case for DNA repair. However, if one wishes to investigate the relevance of the metabolic events to cancer, one must sooner or later turn to mammalian systems. Therefore, we have attempted to purify the enzymes involved in DNA repair from rat liver. A repair endonuclease has been purified 12,000 times to homogeneity on polyacrylamide gel electrophoresis (Biochim. Biophys. Acta 353:99-114, 1974; Biochim. Biophys. Acta 402:343-350, 1975). An exonuclease and a DNA polymerase have been purified 1200 times (J. Cell Biol. 63: 350, 1974). The mammalian endonuclease acts on double-stranded DNA damaged by UV light, or on DNA to which acetylaminofluorene has been bound. It is inactive on single-stranded, or apurinic DNA.

If the activity of the endonuclease is followed by treatment with alkaline phosphatase and DNA polymerase I obtained from bacteria, thymine dimers, or acetylaminofluorene bound to guanosine are released.

The enzyme also acts on X-irradiated DNA, but we don't know what is released after treatment with alkaline phosphatase and DNA polymerase. The enzyme is inhibited by the phorbol esters that are cocarcinogenic. However, it is unlikely that the mechanism of cocarcinogensis can be explained by inhibition of repair of the endonuclease. Phorbol esters that are not cocarcinogenic do not inhibit the enzyme.

A critical question concerns the restrictions of the enzyme activity. Damage to UV irradiated heterochromatin is much more marked than damage to UV irradiated euchromatin. Moreover, heterochromatin is a better substrate for the repair endonuclease than euchromatin. It is, however, debatable whether the endonuclease is rate limiting in repair. In regenerating liver the enzyme activity increases considerably within 2 hours after partial hepatectomy. The increase is due to de novo synthesis. Indeed, it is blocked by actinomycin, and the purified enzymes incorporate leucine-C^{14}, and valine more actively than in normal liver. I thought it to be worthwhile to point out how we have in mammalian studies benefited from the work of Grossman and yourself on bacteria.

D. Goldthwait, Case Western Reserve University: Your enzyme is very interesting particularly in its specificity for both UV dimers and DMBA derivatives. Your enzyme has a more broad substrate specificity than the E. coli enzyme and suggests that what is true for E. coli is not always true for man.

J. Van Lancker: Dr. Bacchetti and Dr. Benne have shown that there is an enzyme in the thymus that has properties similar to ours. Moreover, it doesn't surprise me that in mammalian cells one wouldn't find the variety of repair enzymes that are found in bacteria.

S. Grossman, Union Carbide: I am interested in knowing in mammalian systems in view of the association between chromatin and histones how accessible is the DNA to repair processes?

D. Goldthwait: We have not looked at that, perhaps Dr. Van Lancker could answer that question.

S. Grossman: Well how accessible is DNA to the repair process in mammalian systems in view of the association between DNA and histones?

J. Van Lancker: The studies on euchromatin and heterochromatin extracted from irradiated nuclei were aimed at answering your question. At this point there is no indication that there is a difference in the accessibility of hetero-or euchromatin to the endonuclease. Previous in vivo data on regenerating liver brought us, however, to conclude that at least after the administration of X-radiation derepressed DNA was more rapidly repaired than repressed DNA (Fed. Proc. 29: 1439-1442, 1970).

S. Greer, University of Miami: Can one attribute mutagenesis of the phobolester to its action on the repair enzymes?

D. Goldthwait: I do not know the answer to that. First of all I did not know that phorbolester is mutagenic. Is that what you are saying?

S. Greer: The fact that it would inhibit a repair enzyme, I thought it would be mutagenic. You have not looked into it?

D. Goldthwait: No, I do not know of any evidence for it.

S. Greer: Another possibility is that it might be used for a co-antitumor agent in combination chemotherapy by inhibiting repair of the modification by certain antitumor agents. It might enhance the effectiveness of antitumor activity in combination.

ZINC BIOCHEMISTRY IN THE NORMAL
AND NEOPLASTIC GROWTH PROCESSES

BERT L. VALLEE
Biophysics Research Laboratory
Department of Biological Chemistry
Harvard Medical School
and the
Division of Medical Biology
Peter Bent Brigham Hospital
Boston, Massachusetts

INTRODUCTION

  The presence of metals in biological matter in amounts approaching their detection limits has intrigued biologists for generations, though efforts to ascertain their functional significance have often proved frustrating. The remarkable acceleration of the rate of progress in this field is the result of conjoint advances in many disciplines. Nutritional and metabolic experiments can now be monitored both by advanced methods of analysis and through suitable control of contamination. Major progress in isolating and characterizing the composition, structure and function of metalloenzymes has greatly aided the delineation of the molecular basis of the biological role of metals. Simultaneously, the emerging knowledge has given new direction to experiments in biochemistry, physiology, pathology, nutrition and medicine, and the resultant understanding of metallobiochemistry has given hope that metals may play hitherto unrecognized roles in disease. The possibility that metals might become therapeutic agents has motivated much of the past effort in this field.

  Such considerations are directly pertinent to zinc metabolism. The biological importance of zinc has emerged only during the last two decades. In succession its biological effects were viewed as mostly harmful, questionable and now essential. It is now 100 years ago that zinc was found to be indispensable for the growth of <u>Aspergillus niger</u> (1), and its presence in plants and animals was established within a decade (2,3). Thereafter, despite the rapid growth of biological science during the next half-century, almost no progress was made regarding its biological role which remained conjectural. Not until 1934 was conclusive evidence obtained that zinc is essential to normal growth and development of rodents (4,5).

It is now evident that zinc is essential to normal growth and development of all living matter, but, remarkably, this universal requirement was unappreciated until quite recently (6). Its deficiency results in major abnormalities of composition and function, though the manifestations are complex and can vary depending upon the particular species studied (7-11) (Table I). An increasing number of disease entities are proving to be related to zinc deficiency, both in animals and man (12-16), and this deficiency during pregnancy results in congenital malformations of the embryo particularly by affecting growing or proliferating tissues. The consequences of more subtle metabolic interactions, as in cirrhosis (18), and the basis of genetic or teratological defects (19) have not been examined widely and offer rich investigative opportunities.

Biochemical Approaches to the Study of Zinc

At the end of the last century investigations of cellular respiration and the functional role of iron in oxidative processes first suggested that metals may be essential to enzymatic reactions, pointing the way to the eventual discovery of metalloenzymes. However, the idea that — other than iron — the first transition and IIB group of metals might play significant roles was not appreciated and biochemical knowledge was virtually nonexistent. Experimental work leading to this realization has been performed largely in the last generation (20).

The search for an explanation of the physiologic role of metals has emphasized their interaction with proteins, especially enzymes. The resultant metalloenzymes are characterized by their stability constants, whose magnitudes have served as the operational basis for the two extremes of metal-protein interactions, i.e., the very stable <u>metallo-</u>

enzymes and the more labile metal-enzyme complexes (21), the latter exemplified by sodium, potassium, calcium and magnesium, the most abundant cations of mammalian species.

Owing to their electronic structures, the transition metals and zinc tend to form stable complexes with proteins which exhibit characteristic functions, exemplified by the oxygen-carrying capacity of iron in hemoglobin or of copper in hemocyanin and many enzymes containing these or other metals. The red iron (22) and blue copper proteins (23) early called attention to themselves. It would seem that in large measure the lack of visible color of zinc proteins can be held accountable for the long delay in their recognition. In fact, the isolation, purification and recognition in 1940 (24) of carbonic anhydrase as a zinc enzyme was a happy accident. Through much of their purification, Keilin and Mann's attention was directed to a blue protein, apparently exhibiting carbonic anhydrase activity, and, hence, it was suspected to be a copper enzyme. This blue protein actually proved to be hematocuprein, now recognized to be superoxide dismutase. However, as a result of only minor adjustments of conditions, the analytical method employed for copper could also detect zinc, leading to the recognition of carbonic anhydrase as the first zinc metalloenzyme.

Eighty-five years elapsed between the initial recognition of the metabolic effects of zinc deficiency (1) and the identification of bovine pancreatic carboxypeptidase A as a zinc enzyme in 1954 (25). Additional zinc metalloenzymes were discovered quite rapidly during the next several years (6,21) and, now, after another twenty years, in excess of 70 zinc enzymes have been identified, the majority within the past decade. Zinc is now known to participate in a wide variety of metabolic processes including carbohydrate, lipid, protein and nucleic acid synthesis or degradation. It is essential for the function and/or structure of at least one in each of the six categories of enzymes designated by the Commission on Enzyme Nomenclature of the International Union of Biochemistry and present throughout all phyla (Table II), among them several dehydrogenases, aldolases, peptidases, phosphatases, an isomerase, a transphosphorylase and aspartate transcarbamylase (26).

The detection, among other metals, of substantial quantities of firmly bound zinc in RNA (27) and DNA (28) has revealed additionally important avenues for investigation of its role in biology. Zinc apparently stabilizes the secondary and tertiary structure of RNA (29) and plays an important role in protein synthesis. As yet there is virtually no information on the association of zinc with lipids or carbohydrates, glyco- and lipoproteins, and their systematic analysis might turn out to be revealing.

Zinc proteins and enzymes offer unusual opportunities to study the manner in which specific function is achieved through interaction of the metal with the protein. When zinc can be removed and restored with concomitant loss and restitution of enzymatic activity, differential chemical labeling of the sites of the protein to which the metal binds becomes possible. Additionally, this permits physico-chemical approaches to discern what contribution, if any, zinc makes to overall protein structure and stability. In a number of systems other metals can be substituted successfully for the native zinc, with corresponding characteristic alterations of physical properties and/or catalytic function. These and other approaches have aided in the delineation of the role of zinc in the function and structure of enzymes (20).

Chemical Features of Zinc Enzymes

Zinc, a IIB element with a completed $\underline{d}$ subshell and two additional $\underline{s}$ electrons, chemically combines in the +2 oxidation state. There is no evidence that it is oxidized or reduced in biological reactions. It generally forms tetrahedral complex ions, but many octahedral complexes are known.

The characteristics of zinc in e.g., simple halo-, cyano, and amino-complexes likely differs from those in metalloenzymes, the stereo-chemistry being determined largely by ligand size, electrostatic and covalent binding forces. The three-dimensional structure of proteins, heterogeneity of ligands and the degree of vicinal polarity of the metal binding site may jointly generate atypical coordination properties (30). Unusual bond lengths, distorted geometries, and/or an odd number of ligands can generate a metal binding site on the enzyme, which when occupied by a metal, can be thermodynamically more energetic than metal ions when free in solution where they are complexed to water or simple ligands. As a result, in enzymes zinc is thought to be poised for its intended catalytic function in the entatic state (30). In this context the term entasis indicates the existence of a condition of tension or stress in a zinc — or other metal — enzyme prior to combining with substrate.

In effect, the entatic state is thought to originate in the genetic heritage of the cell. The primary structure of the enzyme protein dictates the relative spatial positions of those amino acid side chains destined to serve as ligands when the apoprotein combines with the metal ion. Evidence suggests that the metal ion is not incorporated into the growing, ribosome-bound polypeptide chain (31), until the protein is fully formed. According to this view, the metal does not induce its own coordination site, but its interaction awaits the expression of the genetic message.

The lack of suitable physical-chemical probe properties of the diamagnetic zinc atom led to a search for means to replace it with paramagnetic metals, e.g., cobalt which could signal information on the nature and environment of the active site (32). The spectra of such cobalt substituted zinc enzymes (33-36) differ significantly from those of model Co^{2+} complexes and are thought to reflect the entatic state of the cobalt (and zinc) ions in these enzymes (30).

Identification of the metal binding ligands of metalloenzymes has been difficult by means other than x-ray crystallography. Cysteinyl, histidyl, tyrosyl residues and the carboxyl groups of aspartic and glutamic acids have been implicated most commonly (20). Thus far, metal complex ions in which the mode of metal coordination is known quite precisely have not proven adequate to define metal binding in enzymes, invariably lacking the entatic environment seemingly characteristic of metalloenzymes (30). Moreover, the complexes which are known most thoroughly and have been studied most extensively are bidentate, but present evidence indicates that in metalloenzymes, metals are more likely coordinated to at least three ligands (20). With few exceptions (37), multidentate complex ions, suitable for appropriate comparisons with metalloenzymes, have not been studied. Yet absorption (38), magnetic circular dichroic (39) and electron paramagnetic resonance spectra (40) combined with kinetic studies of cobalt-substituted and chemically modified zinc metalloenzymes have enlarged understanding both of the possible modes of interaction of zinc with the active sites of zinc metalloenzymes and their potential mechanisms of action.

The molecular details of metalloenzyme action have been elucidated greatly in the past few years (41). Crystal structures for bovine carboxypeptidase A (42), thermolysin (43) and horse liver alcohol dehydrogenase (44) are now available, and chemical and kinetic studies have defined the role of zinc in substrate binding and catalysis (41,20). In fact, many of the significant features elucidating the mode of action of enzymes, in general, have been defined at the hands of zinc metalloenzymes.

Zinc and Normal Cell Biology

In spite of these tremendous advances establishing in but two decades both the very participation of zinc in enzymatic catalysis and many aspects of its presumable mechanisms, knowledge concerning the roles of this element in cellular metabolism is surprisingly sparse and the reactions to which it becomes limiting have not been recognized, defined or integrated. Studies of the role of zinc in cell metabolism requires an organism which can be obtained in homogeneous form, grows rapidly when the trace metal content of the medium is controlled rigorously, and which can be disrupted readily to allow definitive measurements on subcellular organelles. Such an organism, a critical prerequisite for such cell biological studies, has been available only in the last decade: the alga, Euglena gracilis, was found to satisfy the necessary criteria (45). It has now been used as a model to study the biochemical and morphological consequences of zinc deprivation (45-50).

A number of striking chemical changes accompany zinc deficiency-induced growth arrest: RNA and protein synthesis is depressed; cellular DNA content doubles; cell volume increases, and peptides, amino acids, nucleotides, polyphosphates and unusual proteins accumulate. Further, the intracellular content of Ca, Mg, Mn and Fe increases from 6- to 35-fold (49,51). This is particularly interesting since both Mg and Mn activate a number of zinc enzymes, e.g., leucine aminopeptidases, alkaline phosphatases and DNA and RNA polymerases (see below). While the specific biochemical derangements which cause these metabolic and metal imbalances in zinc deficient E. gracilis remained unknown, the evidence was impressive that they would relate to nucleic acid metabolism and cellular division. Reports recognizing DNA and RNA polymerases of prokaryotic organisms as zinc metalloenzymes (52-56) strongly supported this deduction. Recent experiments with zinc deficient E. gracilis, to be summarized briefly, now leave little doubt in this regard and, moreover, detail the effect of the metal on the dynamics of DNA metabolism in the cell cycle.

DNA Metabolism in Zinc Deficient E. Gracilis

When zinc in the medium is sufficient (10^{-5} M), the growth of E. gracilis lags for 5-6 days, followed by a logarithmic growth period continuing up to the 12th-14th days and then enter a stationary phase without further increase in total numbers of cells. But when zinc is limiting (10^{-7} M), as in virtually all other species (Table I), cellular proliferation decreases markedly, most noticeably after a week of incubation (Fig. 1). Solely on addition of zinc, cell growth is resumed within 24-48 hours and reaches precisely the same level as that in initially zinc sufficient cells. Hence, in this instance, growth impairment is due solely to a zinc deficiency. The zinc content of 13-day old cells grown in zinc sufficient media is almost 10-fold greater than that of cells of the same age grown in zinc deficient media, 58 vs 8 µg Zn/10^8 cells.

In addition to chemical and enzymatic alterations (Table I), marked morphological changes are characteristic of the various normal growth phases of E. gracilis (46,47). Cells grown in the presence of zinc are mostly oval during the lag period, but become thin and elongated in the course of the logarithmic and stationary phases and the overall size of the cells decreases, probably due to changes in paramylon content (Fig. 2A). In contrast, cells grown in zinc deficient media remain round or ovoid throughout the growth period. Their size increases continuously from the lag through the log and stationary phases, unlike that of the controls (Fig. 2B).

Electron microscopy reveals surprisingly little change either in size or in ultrastructure of the principal cellular organelles of zinc deficient cells (47). The abundance of ribosomes may be diminished but the mitochondria, Golgi bodies and endoplasmic reticulum seem unaltered. Importantly there is no change in nuclear morphology. In fact, the extraordinary accumulation of paramylon, the storage form of carbohydrate in these organisms, represents the most striking change (Fig. 2).

We have examined the events during the typical, eukaryotic cell cycle in order to better define the specific lesions that accompany zinc deficiency. Though E. gracilis is a eukaryote, it undergoes growth phases which are usually associated with prokaryotic organisms, i.e., lag, log and stationary phases. Yet this division is best characterized by detailed phases of G_1, S, G_2 and mitosis, as described by Howard and Pelc (57). In these terms, E. gracilis cells increase in size and synthesize RNA and protein during G_1. In S, increased DNA synthesis results in chromosomal replication. The mitotic apparatus forms during G_2 followed by nuclear and cell division.

Potentially, zinc deprivation affects all of these steps (Table III). In E. gracilis zinc might similarly be involved in nucleic-acid metabolism and mitotic activity and its deficiency might affect multiple processes all of which bear on cell division and nucleic acid function.

In order to detail, delineate and better define those steps in the cell cycle of E. gracilis which zinc deprivation affects specifically, we have examined samples of cells at each phase of this sequence of cell division by means of the laser excited cytofluorometer (48). The growth of E. gracilis can be synchronized by exposing the cells to alternating periods of light, which retard cell division, and darkness, during which cells undergo mitosis (Fig. 3). In the light period, at various known intervals, i.e., 4, 8 and 14 hours, cells pass through G_1, S and G_2 (58). Each of these phases can then be examined for their DNA content by means of flow cytofluorometry (48, 59-63).

Aliquots of synchronized zinc sufficient cells for analysis of DNA content were collected at different times by centrifugation, fixed in ethanol and stained with propidium diiodide after treatment with ribonuclease to remove RNA. The stained cells were analyzed in a cytofluorograph provided with a cell sorter which plots the results as histograms of the numbers of cells vs. increasing values of cellular fluorescence, constituting a direct measure of DNA content (Fig. 4). The pattern of cells harvested after 4 hours of exposure to light represents a single complement of DNA typical of cells in G_1 (Fig. 5, panel A). Fewer numbers of cells harvested after 8 hours (Fig. 5, panel B) are in G_1, whereas the number of cells which are synthesizing DNA, i.e., those in S phase, increase. After 14 hours most of the cells have reached G_2, as evidenced by a doubling of their DNA content and a corresponding shift in the fluorescence pattern (Fig. 5, panel C).

The pattern obtained for zinc sufficient cells, taken at early stationary phase (Fig. 6A) when net cell division ceases, indicates that most of the cells are in G_1 with the balance in S. This pattern contrasts with that obtained for zinc deficient cells recorded also at a time when further growth does not occur (Fig. 6B). These cells are predominantly at the S/G_2 interface suggesting that, as these nonsynchronously growing cells are deprived of zinc, those that are in S do not continue into G_2, while those in G_2 do not proceed through mitosis.

Transfer of zinc sufficient cells in early stationary phase, known to be mostly in G_1, to zinc deficient medium reveals further details of the effect of zinc deficiency on the G_1 to S transition. Initially, the number of cells increases by 25%, but growth soon ceases, though it can be restored by addition of zinc which causes a 200% increase in the number of cells within 24 hours. Hence, zinc deprivation of cells in G_1 blocks their progression into S, which zinc then restores. Clearly, the biochemical processes essential for cells to pass from G_1 into S, from S to G_2 and from G_2 to mitosis require zinc, and its deficiency can block all phases of the growth cycle of E. gracilis.

Zinc is an enzymatically essential component of both DNA-dependent DNA and RNA polymerases isolated from prokaryotic organisms (55,56) but its role in the corresponding enzymes from eukaryotes has not been detailed. Its probable requirement by these organisms is illustrated by the reduced incorporation of labeled DNA and RNA precursors into the livers of zinc-deficient rats (64), cultured chick embryos (65) and phytohemagglutinin-stimulated lymphocytes (66). Further, the rate of incorporation of ^{3}H-uridine into RNA of zinc deficient E. gracilis is decreased (47), the element maintains the integrity of ribosomes of these organisms (67) and seems required for the stability of both RNA and DNA (28,62). Most recently, zinc has been demonstrated to be essential to the function of protein synthesis elongation factor 1 (Light Form EF_L) from rat liver which catalyzes the binding of aminoacyl-t RNA to an RNA-ribosome site through the formation of an aminoacyl-t RNA·EF1·GTP ternary complex (68).

The distinctive red complex of dithizone with zinc (69) is the basis of one of the most sensitive chemical means to quantitate this metal. This reaction has served to demonstrate the presence of zinc histochemically in the nucleus and in subnuclear structures and its translocation to and from the nucleolus, spindle apparatus and chromosomes at each stage of mitosis. Zinc dithizonate complexes are observed in the nucleolus and spindle apparatus during prophase, then in chromosomes during metaphase and anaphase, followed by a reappearance in the nucleolus in telophase (70). Thus, there is evidence for the participation of zinc in nucleic acid synthesis, protein synthesis and cell division entirely consistent with the results of the above cell cytofluorometry.

The expectation that in E. gracilis zinc might be essential to systems involved in the polymerization of RNA and/or DNA would seem reasonable, constituting biochemical foundations of the above observations — though the metal is apparently indispensable at multiple loci. Three years ago, as a part of a systematic effort to unravel these interrelations, Dr. Kenneth Falchuk undertook to study the nature and role of these enzymes in zinc sufficient E. gracilis.

DNA Dependent RNA Polymerase II of Euglena gracilis, A Zinc Metalloenzyme

The isolation of DNA dependent RNA polymerase from E. gracilis proved difficult, but the problems encountered in purifying this enzyme were entirely analogous to those described for its isolation from other eukaryotic organisms. Yet, examination of a possible role of zinc in the function of RNA polymerase clearly depends on its purification to homogeneity in quantities sufficient for metal analysis. While the technical problems are demanding, decisive advances both in methods of isolation of eukaryotic polymerases (71) and metal analysis, e.g., microwave excitation emission spectroscopy (72) during the last two years allowed such characterization of the enzyme. The spectroscopic procedure accurately determines picogram quantities of metal ions in microgram quantities of this and other polymerases (72) (see below) demonstrating that it is a zinc enzyme (73).

RNA polymerases from E. gracilis were solubilized by cell disruption in high salt buffers (74,75) and resolved into RNA polymerases I and II, based on their elution patterns from DEAE Sephadex chromatography. The enzymes are completely dependent on exogenous DNA and on all four nucleotides for activity. In both instances, the product of the complete reaction is sensitive to RNAse — but not to DNAse — digestion. RNA polymerase I is insensitive and polymerase II is sensitive to α-amanitine. The pH optimum of RNA polymerase II is pH 7.9 in 0.1 M Tris, with the activity varying as a function of the source and physical state of the template. The two enzymes meet all criteria established so far for the identification of DNA dependent RNA polymerases.

RNA polymerase II has been purified to homogeneity and has a provisional molecular weight of approximately 700,000 daltons. It is composed of multiple subunits of varied though as yet unknown molecular weights, the determination of which is essential to define the molecular weight of the holoenzyme precisely and, hence, to derive an accurate metal/protein ratio. Keeping in mind the molecular weight as the basis of any uncertainty regarding the ratio, the E. gracilis RNA polymerase II contains an average of 2.2 g-atom of zinc per molecular weight 700,000; Mn, Cu and Fe were not found (Table IV). The enzyme is inhibited by chelating agents, e.g., 8-hydroxyquinoline, 8-hydroxyquinoline-5-sulfonic acid, EDTA, α-α'bipyridyl, lomofungin and 1,10-phenanthroline, but not its nonchelating isomers (see Table V). The inhibition is instantaneous, fully reversible by dilution and by addition of an excess of metal ions; it is competitive with nucleotide triphosphate substrates but not with templates. Similar studies of the RNA polymerase I from E. gracilis are presently in progress. In this regard it is of great interest that D.S. Auld and P. Valenzuela (76) have just found the eukaryotic DNA dependent RNA polymerase I of yeast to be a zinc enzyme containing 2.4 g-atom of Zn per molecular weight 650,000.

The isolation of the DNA dependent DNA polymerases of E. gracilis and evidence of inhibition by 1,10-phenanthroline (77,78) rounds out the rapidly emerging evidence for the critical role of zinc in the nucleic acid polymerases of E. gracilis and — presumably — other eukaryotic organisms. It is apparent that multiple processes, critical to the development, division and differentiation of cells, are zinc dependent as shown by the cytofluorometric studies, confirmation that both DNA and RNA polymerases are zinc dependent which calls for experiments which can resolve whether or not interference with the functions of these enzymes becomes limiting to growth in zinc deficiency.

We have noted that — in a given zinc enzyme — the metal can either be indispensable to activity, stabilize its structure or regulate its activity or any combination of these (79-81). It is conjectural at present which enzymatic step and role of the metal will prove to be growth limiting in zinc deficiency — if, indeed, it is an enzymatic step that will turn out to be accountable. It would seem apparent that the rapidly accumulating results of metallobiochemical investigations will soon reveal the basis of hitherto empirical observations of cell biological studies.

In this regard it will be recalled that in zinc deficient E. gracilis, Mg, Mn, Ca, Fe and perhaps other metals accumulate (47,51). Mg and Mn have previously been ascribed roles in substrate, template or inhibitor binding of nucleic acid polymerases (55,56,82,83), and the enzyme can be virtually inactive in their absence. It is conceivable that this accumulation reflects a metabolic response designed to compensate for the effects of zinc deficiency through the formation of metallopolymerases in which other metals substitute for zinc. These, in turn, might exhibit different specificities than the corresponding native zinc RNA (or DNA) polymerases (20). Such conjectures would seem consistent with previous findings, demonstrating that base composition of RNA products is altered when RNA polymerase is exposed to Mg and Mn, respectively (82). The synthesis of abnormal products by zinc deficient E. gracilis (51) would be consistent with such a postulate, which can be tested, among others, by isolation and characterization of the RNA and DNA polymerases from zinc deficient cells.

Zinc and Neoplastic Disease

In spite of the now extensive biological literature documenting profound effects of zinc on normal growth and development, its possible role in abnormal growth has been examined in much more cursory fashion. The demonstration that a series of reverse transcriptases are zinc metalloenzymes has unexpectedly linked earlier observations of a role of zinc in abnormal growth to present day enzymology.

Reports that many human and other vertebrate contain zinc (84) thought to be particularly abundant in certain tumors (85-87),first generated interest in the possible role of this element in neoplasia. These and other early reports of abnormal concentrations of this element in tissues or fluids of tumor bearing animals may now be primarily of historical interest, considering the then available methodology.

Teratomata

The role of zinc in carcinogenesis was explored by injecting zinc chloride or sulfate into the cock testis generating teratomata which can be transplanted (88,89); zinc acetate or zinc stearate proved ineffective (90). These seasonal tumors develop only when the injections are given between January and March, periods of high sexual activity (89,90). Spontaneous teratomata are rare in this species.

Leukemia

A number of years ago we found substantial quantities of zinc in normal human leukocytes (91), later shown to be associated largely with two zinc proteins whose functional identities remain to be established (92,93).

Such studies of the zinc metabolism of normal and leukemic leukocytes generated the postulate that disturbance of a zinc dependent enzyme might be critical to the pathophysiology of myelogenous and lymphatic leukemia, to quote: "Since zinc is known to be present in carbonic anhydrase, it is quite possible that there is another enzyme system with which it is concerned in myelopoiesis, and that there is some disturbance of this enzyme in leukemia" (94,95).

This hypothesis has recently found support in studies of the Type C oncogenic RNA viruses, e.g., avian myeloblastosis virus, which are associated with lymphomas and leukemia in a number of species. The existence of an RNA-dependent DNA polymerase — reverse transcriptase — in these RNA tumor viruses (96,97) has greatly stimulated study of the initiation, biochemical basis and maintenance of malignant transformations and of the manner by which viral RNA is transcribed into a DNA copy. Although a role for zinc in this process was unknown, the above indications of its importance in normal and leukemic leukocyte metabolism prompted us to examine the zinc content of the RNA-dependent DNA polymerases from avian myeloblastosis, murine, simian, feline and RD-114 RNA tumor viruses. Complexing agents have long been employed to explore the functional role of metals in enzymes by kinetic methods. However, even if a series of these agents are employed to study their inhibition of enzymatic activity, of themselves, their effects do not allow a definitive decision regarding the specific metal atom which may be involved in a particular instance (20). The combination of results obtained with agents which differ in regard to their selectivity can be suggestive, but ultimately, the identification of a metalloenzyme minimally requires analytical demonstration of the presence of a functional metal atom. The paucity of the material seemed to present formidable problems in regard to metal analysis, but the marked inhibition of the enzyme's activity by metal-binding agents led us to develop and employ an instrument for microwave excitation spectroscopy (Fig. 7), a means capable of quantitative metal determinations at the 10^{-14} g-atom level. This allowed the identification of stoichiometric amounts of zinc in these enzymes which was available to us in quantitities of the order of 10^{-9} moles. The limits of detection of the procedures are sufficiently low, ultimately to permit quantitative studies of metals and their metabolism in microgram quantities of enzyme (52,53,72). The enzyme employed exhibited activities comparable to those of the most highly purified preparations then reported.

Using optimal conditions for the assay (53), the K_m for TTP was 1×10^{-5} M and the V_{max} was 12 pmoles TMP incorporated per minute per μg of protein at 25°.

Metal complexing agents inhibit the AMV polymerase catalyzed reaction. In their absence, the enzyme is completely stable and active for 60 minutes at 25° while in the presence of 1,10-phenanthroline, 1 mM, the enzyme is inhibited instantaneously and reversibly with a K_I of 7×10^{-5} M.

The metal complexing properties of 1,10-phenanthroline account for the inhibition: its isomers 1,7- and 4,7-phenanthroline neither bind metals nor do they inhibit the nucleotide polymerization reaction under conditions where 1,10-phenanthroline inhibits it completely. Moreover, a number of other, structurally different metal binding agents, e.g., EDTA and 8-hydroxyquinoline-5-sulfonate also markedly inhibit the polymerase activity instantaneously (Table V). The presence of stoichiometric quantities of metal is, of course, essential to verify that these agents exert their effect by binding to a functional and/or structurally essential metal.

Conventional procedures have not been sufficiently sensitive to allow quantitative metal determinations on very small amounts of enzyme. Microwave induced emission spectrometry, however, can extend the usual detection limits by 6 orders of magnitude to 10^{-14} g-atom. Thus, precise quantitative metal analyses on the microgram amounts of enzyme available to us for this purpose seemed feasible by this means. Zn, Cu or Fe could account for the observed inhibition of the AMV polymerase, and these elements and Mn were determined after removal of metal quenching agents and low molecular weight protein contaminants by gel exclusion chromatography. The elements and protein were measured quantitatively with high precision in 45 µl fractions containing maximal activity when absolute amounts of metal varied from 10^{-11} to 10^{-14} g-atom. The purified enzyme contains zinc in stoichiometric quantities but Cu, Fe, and Mn are virtually absent (Table VI). The zinc content corresponds to from 1.7 to 1.9 g-atom of zinc per mole of enzyme for molecular weights of either 1.6 to 1.8 x 10^5.

In similar fashion, the murine, simian, feline and RD-114 Type C RNA tumor viruses have been shown to be zinc enzymes by measuring microgram quantities of zinc in these proteins. All of them are inhibited by 1,10-phenanthroline (Table V) and all contain stoichiometric quantities of zinc, assuming specific activities similar to that of the enzyme from avian myeloblastosis virus.

These data extend the role of zinc in enzymes essential to normal nucleic acid metabolism to others presumed to play a role in a leukemic process, confirming a hypothesis of long standing (94,95). This, in turn, led us to examine the role of zinc in nucleic acid metabolism of leukemic leukocytes.

1,10-Phenanthroline Dependent Growth Inhibition of Human Lymphoblasts

We have now examined the DNA metabolism of CCRF-CEM lymphoblasts[*] grown in cell culture, employing instrumentation and procedures described above in regard to the flow cytofluorometric examination of zinc sufficient and deficient <u>Euglena Gracilis</u> (Fig. IV). Hydroxyurea results in a block of division of these cells, but their subsequent incubation in fresh media releases it and synchronizes their

---

[*]This human lymphoid cell line was established by cultivating cells from the blood of leukemic, pediatric patients.

division with a subsequent 100% increase in the number of cells in 12 hours. By releasing the block, at discrete time intervals the DNA metabolism of synchronized cells can be followed by means of flow cytofluorometry performed at specific stages of the cell cycle. These are identified and defined by the extent to which ^{3}H thymidine is incorporated and by the resultant labeling and mitotic indices. In particular, the DNA content of lymphoblasts in G_2 (61) can be ascertained by blocking with podophyllotoxin. Lymphoblasts have not been grown under zinc deficient conditions thus far, but exposure to 1,10-phenanthroline can serve to mimic this condition. This agent forms stable zinc complexes, both when zinc is free in solution and when it is firmly incorporated into proteins and enzymes. As indicated earlier, 1,10-phenanthroline inhibits zinc dependent reactions and the consequences can closely resemble those of zinc deprivation, though this agent interacts with other metals of the first transition group as well, of course.

In normal media, the number of lymphoblasts double within 24 hours but 1,10-phenanthroline, 4 µM, completely arrests their proliferation and higher concentrations are cytotoxic, actually decreasing their number by 20% compared with the control. This contrasts markedly with the consequences of exposure to the nonchelating isomer, 1,7-phenanthroline, which fails to affect proliferation at any concentration employed, suggesting that 1,10-phenanthroline exerts its action by chelating a metal essential to cell division. Both dilution and the addition to the medium of Zn^{2+}, Cu^{2+}, or Fe^{2+} results in the resumption of proliferation in a manner reminiscent of the reversal of 1,10-phenanthroline inhibition of specific enzymes (20).

The synchronization of CEM lymphoblasts, either blocked with hydroxyurea or with podophyllotoxin, and their subsequent release from the block at specified time intervals synchronizes cells at specific stages of the cell cycle. Subsequent proliferation in the presence of 1,10-phenanthroline can then be correlated to the distribution and content of the DNA of the resultant cell population and this permits localization of its effects to a particular stage of cell division. 1,10-Phenanthroline inhibits cell division of lymphoblasts by blocking their transition from the G_1 into the S phase. Further, lymphoblasts which are synthesizing DNA in the S phase are blocked and do not progress into G_2 (98,99).

Inhibition of ^{3}H thymidine incorporation suggests that the failure to progress from G_1 into S must involve processes essential for the synthesis of DNA itself. Such inhibition of ^{3}H thymidine incorporation has also been observed with phytohemagglutinin stimulated peripheral lymphocytes incubated with 1,10-phenanthroline (66), chicken embryo (65) and kidney cortex cells incubated with EDTA, all suggesting that — both in mammalian and nonmammalian cells — metal-dependent processes are critical to the transition from G_1 to S and to the progression through S itself. The metabolic properties of both phases are characteristic. In G_1 the activity of enzymes and uptake of substrates essential for synthesis, e.g., amino acids, nucleotides, metals, etc., all increase. DNA synthesis occurs in S. Thus, in these two stages, cells synthesize DNA, RNA and proteins, essential for cell division, and the 1,10-phenanthroline data indicate metal-dependent processes to be most critical (98,99).

Such conclusions are consistent with earlier studies both of histochemical localization of zinc in a number of systems and the consequences of its deficiency. It is essential to the incorporation of labeled DNA and/or RNA precursors into rat embryos (100) and livers of zinc deficient rats (64). It has been localized to the nucleoli of sea urchin eggs and, transiently, to the mitotic apparatus of dividing cells (70), suggesting a role in mitosis. Finally, 1,10-phenanthroline,<4 µM, completely arrests proliferation both of normal and leukemic CEM lymphoblasts but, strikingly, >4µM destroys only the leukemic CEM cells. The differentiation of these two types of cells, based on their response to 1,10-phenanthroline, may perhaps relate to the known difference of their zinc contents and could suggest novel therapeutic approaches to the disease.

While the present studies do not permit an allocation of these effects either to specific zinc systems or, in fact, to specific biochemical processes involved in the cell cycle, they clearly do indicate that derangements of metal metabolism, likely zinc metabolism, exerts its effect in G_1 and S, premitotic phases of the cell cycle of leukemic cells.

EPILOG

The ubiquitous occurrence of zinc in subcellular organelles and in enzymes, critical both to protein and nucleic acid metabolism, calls for far more detailed examination of its role in neoplastic growth processes. Present indications are that the study of zinc metabolism may well provide a novel and rewarding focus for the examination of this puzzling group of derangements.

REFERENCES

(1) J. Raulin, Ann. Sci. Natl. Botan. et Biol. Vegetale, 11 (1869) 93.

(2) G. Lechartier and F. Bellamy, Compt. Rend. Acad. Sci., 84 (1877) 687.

(3) F. Raoult and H. Breton, Compt. Rend. Acad. Sci., 85 (1877) 40.

(4) G. Bertrand and R.C. Bhattacherjee, Compt. Rend. Acad. Sci., 198 (1934) 1823.

(5) W.R. Todd, C.A. Elvehjem and E.G. Hart, Amer. J. Physiol., 107 (1934) 146.

(6) B.L. Vallee, Physiol. Rev., 39 (1959) 443.

(7) B.L. Vallee, J. Chronic Diseases, 9 (1959) 74.

(8) T.-K. Li and B.L. Vallee, in: Modern Nutrition in Health and Disease, 5th Ed., eds. R.S. Goodhart and M.E. Shils (Lea & Febiger, Philadelphia, 1973) p. 372.

(9) E.J. Underwood, Trace Elements in Human and Animal Nutrition, 3rd Ed. (Academic Press, New York, 1971).

(10) A.S. Prasad, Ed. Zinc Metabolism (C.C. Thomas, Springfield, Ill., 1966).

(11) W.J. Pories and W.H. Strain, Eds. Clinical Applications of Zinc Metabolism (C.C. Thomas, Springfield, Ill. 1974)

(12) R.E. Burch, H.K.J. Hahn and J.R. Sullivan, Clin. Chem., 21 (1975) 501.

(13) J.A. Halsted, J.C. Smith, Jr. and M.I. Irwin, J. Nutr., 104 (1974) 347.

(14) E.J. Moynahan, Lancet, 2 (1974) 399.

(15) K.H. Neldner and K.M. Hambidge, New Engl. J. Med., 292 (1975) 879.

(16) H.F. Tucker and W.D. Salmon, Proc. Soc. Exp. Biol. Med., 88 (1955) 613.

(17) L.S. Hurley, Am. J. Nutr., 22 (1969) 1332.

(18) B.L. Vallee, W.E.C. Wacker, A.F. Bartholomay and E.D. Robin, New Engl. J. Med., 255 (1956) 403.

(19) L.S. Hurley and R.E. Shrader, in: Neurobiology of the Trace Metals Zinc and Copper, ed. C.C. Pfeiffer (Academic Press, New York, 1972) p. 7.

(20) B.L. Vallee and W.E.C. Wacker, in: The Proteins, Vol. 5, 2nd Ed., ed. H. Neurath (Academic Press, New York, 1970).

(21) B.L. Vallee, Advances in Protein Chemistry 10 (1955) 317.

(22) W.T. Preyer, De Haemoglobino Observationes et Experimenta (M. Cohen and Son, Bonn, 1866) p. 27.

(23) E. Harless, Müller's Arch, f. Anat., Physiol., 148 (1847)

(24) D. Keilin and T. Mann, Biochem. J., 34 (1940) 1163.

(25) B.L. Vallee and H. Neurath, J. Biol. Chem., 217 (1955) 253.

(26) B.L. Vallee and W.E.C. Wacker, in: C.R.C. Handbook of Biochemistry and Molecular Biology, ed. G. Fasman (C.R.C. Press) in press.

(27) W.E.C. Wacker and B.L. Vallee, J. Biol. Chem., 234 (1959) 3257.

(28) Y.A. Shin and G.L. Eichhorn, J. Amer. Chem. Soc., 90 (1968) 7323.

(29) K. Fuwa, W.E.C. Wacker, R. Druyan, A.F. Bartholomay and B.L. Vallee, Proc. Natl. Acad. Sci. USA, 46 (1960) 1298.

(30) B.L. Vallee and R.J.P. Williams, Proc. Natl. Acad. Sci. USA, 59 (1968) 498.

(31) M.I. Harris and J.E. Coleman, J. Biol. Chem., 243 (1968) 5063.

(32) B.L. Vallee, in: Metal Ions in Biological Systems, Vol. 40, ed. S.K. Dhar (Plenum, New York, 1973) p. 1.

(33) B.L. Vallee and S.A. Latt, in: Structure-Function Relationships of Proteolytic Enzymes, eds. P. Desnuelle, H. Neurath and M. Ottesen (Academic Press, New York, 1970) p. 144.

(34) S. Lindskog and P.O. Nyman, Biochim. Biophys. Acta, 85 (1964) 141.

(35) B. Holmquist, T.A. Kaden and B.L. Vallee, Biochemistry, 14 (1975) 1454.

(36) R.T. Simpson and B.L. Vallee, Biochemistry, 7 (1968) 4343.

(37) T. Herskovitz, B.A. Averill, R.H. Holm, J.A. Ibers, W.D. Phillips and J.F. Weiher, Proc. Natl. Acad. Sci. USA, 69 (1972) 2437.

(38) S.A. Latt and B.L. Vallee, Biochemistry, 10 (1971) 4263.

(39) T.A. Kaden, B. Holmquist and B.L. Vallee, Inorg. Chem., 13 (1974) 2585.

(40) F.S. Kennedy, H.A.O. Hill, T.A. Kaden and B.L. Vallee, Biochem. Biophys. Res. Commun., 48 (1972) 1533.

(41) J.F. Riordan and B.L. Vallee, Advan. Exp. Med. Biol., 48 (1974) 33.

(42) F.A. Quiocho and W.N. Lipscomb, Advan. Protein Chem., 25 (1971) 1.

(43) P.M. Colman, J.N. Jansonius and B.W. Matthews, J. Mol. Biol., 70 (1972) 701.

(44) C.-I. Brandén, H. Jörnvall, H. Eklund and B. Furugren, in: The Enzymes, ed. P.D. Boyer (Academic Press, New York, 1975) p. 104.

(45) C.A. Price and B.L. Vallee, Plant Physiol., 37 (1962) 428.

(46) K.H. Falchuk and D.W. Fawcett, Federation Proc., 33 (1974) 1475.

(47) K.H. Falchuk, D.W. Fawcett and B.L. Vallee, J. Cell Sci., 17 (1975) 57.

(48) K.H. Falchuk, A. Krishan and B.L. Vallee, Biochemistry, 14 (1975) 3439.

(49) W.E.C. Wacker, Biochemistry, 1 (1962) 859.

(50) C.A. Price, in: Zinc Metabolism, ed. A.S. Prasad (C.C. Thomas, Springfield, Ill., 1966) p. 69.

(51) W.E.C. Wacker, W. Kornicker and L. Pothier, Abstr. Amer. Chem. Soc. 150th Meeting (1965) 88C.

(52) D.S. Auld, H. Kawaguchi, D. Livingston and B.L. Vallee, Biochem. Biophys. Res. Commun., 57 (1974) 967.

(53) D.S. Auld, H. Kawaguchi, D. Livingston and B.L. Vallee, Proc. Natl. Acad. Sci. USA, 71 (1974) 2091.

(54) D.S. Auld, H. Kawaguchi, D.M. Livingston and B.L. Vallee, Biochem. Biophys. Res. Commun., 62 (1975) 296.

(55) M.C. Scrutton, C.W. Wu and D.A. Goldthwait, Proc. Natl. Acad. Sci. USA, 68 (1971) 2497.

(56) J.P. Slater, A.S. Mildvan and L.A. Loeb, Biochem. Biophys. Res. Commun., 44 (1971) 37.

(57) A. Howard and S.R. Pelc, Symposium on Chromosome Breakage (supplement to Heredity, Vol. 6) 1953.

(58) L.N. Edmunds, Jr., Science, 145 (1964) 266.

(59) H.A. Crissman and J.A. Steinkamp, J. Cell Biol., 59 (1973) 766.

(60) P.M. Kraemer, L.L. Deaven, H.A. Crissman and M.A. Van Dilla, Advances Cell Molec. Biol., 2 (1972) 47.

(61) A. Krishan, J. Cell Biol., 66 (1975) 521.

(62) R.A. Tobey and H.A. Crissman, Can. Res., 32 (1972) 2726.

(63) M.A. Van Dilla, T.T. Trujillo, P.F. Mullaney and J.R. Coulter, Science, 163 (1969) 1213.

(64) H.H. Sandstead and R.A. Rinaldi, J. Cell. Physiol., 73 (1969) 81.

(65) H. Rubin, Proc. Natl. Acad. Sci. USA, 69 (1972) 712.

(66) R.O. Williams and L. Loeb, J. Cell Biol., 58 (1973) 594.

(67) J.A. Prask and D.J. Plocke, Plant Physiol., 48 (1971) 150.

(68) S. Kotsiopoulos and S.C. Mohr, Biochem. Biophys. Res. Commun., 67 (1975) 979.

(69) B.L. Vallee and J.G. Gibson, J. Biol. Chem., 176 (1948) 435.

(70) T. Fujii, Nature, 174 (1954) 1108.

(71) Cold Spring Harbor Symposia on Quantitative Biology, Vol. 38 (1973).

(72) H. Kawaguchi and B.L. Vallee, Anal. Chem., 47 (1975) 1029.

(73) K.H. Falchuk, B. Mazus, D. Ulpino and B.L. Vallee, in preparation.

(74) R.G. Roeder and W.J. Rutter, Nature, 224 (1969) 234.

(75) P. Chambon, in: The Enzymes, Vol. X, ed. P.D. Boyer (Academic Press, New York, 1974) p. 261.

(76) D.S. Auld and P. Valenzuela, Biochem. Biopnys. Res. Commun., in press.

(77) A.G. McLennan and H.M. Keir, Biochem. J., 151 (1975) 227.

(78) A.G. McLennan and H.M. Keir, Biochem. J., 151 (1975) 239.

(79) J.P. Rosenbusch and K. Weber, Proc. Natl. Acad. Sci. USA, 68 (1971) 68.

(80) E.R. Stadtman, B.M. Shapiro, A. Ginsburg, D.S. Kingdon and M.D. Denton, Brookhaven Symp. Biol., 21 (1968) 378.

(81) R.A. Anderson, W.F. Bosron, F.S. Kennedy and B.L. Vallee, Proc. Natl. Acad. Sci. USA, 72 (1975) 2989.

(82) M. Chamberlin and P. Berg, Proc. Natl. Acad. Sci. USA, 48 (1962) 81.

(83) L.M.S. Chang and F.J. Bollum, Proc. Natl. Acad. Sci. USA, 65 (1970) 1041.

(84) C. Delezenne, Ann. Inst. Pasteur, 33 (1919) 68.

(85) P. Cristol, Contribution a l'étude de la physio-pathologie du zinc et en particulier de sa signification dans les tumeurs. Thèse Montpellier (1922)

(86) P. Cristol, Bull. Soc. Chim. Biol., 5 (1923) 23.

(87) H. Labbé and P. Nepveux, le Progrès Med. (1927) 577.

(88) I. Michailowsky, Virchow's Archiv. Pathol. Anat., 267 (1928) 27.

(89) L.I. Falin and K.E. Gromzewa, Amer. J. Cancer, 36 (1939) 233.

(90) J. Guthrie, Brit. J. Cancer, 18 (1964) 130.

(91) B.L. Vallee and J.G. Gibson, 2nd, J. Biol. Chem., 176 (1948) 445.

(92) F.L. Hoch and B.L. Vallee, J. Biol. Chem., 195 (1952) 531.

(93) B.L. Vallee, F.L. Hoch and W.L. Hughes, Jr., Arch. Biochem. Biophys., 48 (1954) 347.

(94) J.C. Gibson, 2nd, B.L. Vallee, R.G. Fluharty and J.E. Nelson, Acta contre Cancer, 6 (1950) 102.

(95) B.L. Vallee, R.G. Fluharty and J.G. Gibson, 2nd, Acta Union Intern. Contre Cancer, 6 (1949) 869.

(96) H.M. Temin and S. Mizutani, Nature, 226 (1970) 1211.

(97) D. Baltimore, Nature, 226 (1970) 1209.

(98) K.H. Falchuk and A. Krishan, Federation Proc., 34 (1975) 530.

(99) K.H. Falchuk, A. Krishan and B.L. Vallee, in preparation.

(100) H. Swenerton, R. Shader and L.S. Hurley, Science, 166 (1969) 1014.

(101) B.J. Poiesz, N. Battula and L.A. Loeb, Biochem. Biophys. Res. Commun., 56 (1974) 959.

(102) B.J. Poiesz, G. Seal and L.A. Loeb, Proc. Natl. Acad. Sci. USA, 71 (1974) 4892.

Table I. ZINC DEFICIENCY IN DIFFERENT PHYLA

	Growth	Development	Changes in Chemical Composition		Decreased Enzymatic Activities
			Decrease	Increase	
Micro-organisms	Retardation	Increase in cellular size	Protein RNA (ribosomal) Pyridine Nucleotides	DNA Amino acids Polyphosphates Phospholipids ATP Organic acids	Alkaline phosphatase Alcohol and D-Lactate dehydrogenases Tryptophan desmolase
Plants	Retardation	Small abnormal leaves Chlorotic mottling Decreased fruit production	Protein Auxin Ethanolamine	Amino acids	Tryptophan desmolase Carbonic anhydrase Aldolase Pyruvic carboxylase
Vertebrates	Retardation	Testicular atrophy Parakeratosis Dermatitis Coarse, sparse hair	Red blood cells Serum proteins	Uric acid	Alkaline phosphatase Pancreatic proteases Malate, lactic alcohol dehydrogenases NADH diaphorase

TABLE II
Some Zinc Metalloenzymes

Enzyme	IUB no.	Source
Alcohol dehydrogenase	1.1.1.1	Yeast; horse, human liver
D-Lactate cytochrome reductase	1.1.2.4	Yeast
Glyceraldehyde-phosphate dehydrogenase	1.2.1.13	Beef, pig muscle
Phosphoglucomutase	2.7.5.1	Yeast
RNA polymerase	2.7.7.6	E. coli
DNA polymerase	2.7.7.7	E. coli
Reverse transcriptase	2.7.7.-	Avian myeloblastosis virus
Mercaptopyruvate sulfur transferase	2.8.1.2	E. coli
Alkaline phosphatase	3.1.3.1	E. coli
Phospholipase C	3.1.4.3	Bacillus cereus
Leucine aminopeptidase	3.4.1.1	Pig kidney, lens
Carboxypeptidase A	3.4.2.1	Beef, human pancreas
Carboxypeptidase B	3.4.2.2	Beef, pig pancreas
Carboxypeptidase G	3.4.2.-	Pseudomonas stutzeri
Dipeptidase	3.4.3.-	Pig kidney
Neutral protease	3.4.4.-	Bacillus sp.
Alkaline protease	3.4.4.-	Escherichia freundii
AMP aminohydrolase	3.5.4.6	Rabbit muscle
Aldolase	4.1.2.13	Yeast; Aspergillus niger
Carbonic anhydrase	4.2.1.1	Erythrocytes
δ-Aminolevulinic acid dehydratase	4.2.1.24	Beef liver
Phosphomannose isomerase	5.3.1.8	Yeast
Pyruvate carboxylase	6.4.1.1	Yeast

Table III

The Role of Zinc in the Biochemical
and Morphological Events of the Cell Cycle

Cell Cycle	Biochemical and Morphological Events	Processes or Cellular Components With Known Zn Requirement
G_1	RNA Synthesis Protein Synthesis Increase in Size	Uridine Incorporation into RNA DNA-dependent RNA Polymerase RNA Stabilization
S	DNA Synthesis	Thymidine Incorporation into DNA DNA-dependent DNA Polymerase DNA Stabilization
G_2	Assembly of Functioning Apparatus for Nuclear and Cell Division	Unknown
Mitosis	Nuclear Division	Dithizone Staining of Nucleoli Spindle Apparatus Chromosomes

Table IV

E. Gracilis RNA Polymerase II:
Metal Content

Metal	Protein mg/ml	Metal µg/mg prot.	g-atom M.W. 700,000
Zn	0.10	0.204	2.2
	0.10	0.185	2.0
	0.18	0.203	2.2
	0.45	0.251	2.6
Mn	0.10	0	0
Cu	0.10	0	0
Fe	0.10	0	0

Table V

Effect of Metal Binding Agents on the (AMV) RNA Dependent DNA Polymerase Activity[*][**]

Metal Binding Agent	Concentration x 10^4, M	v/v_c[***] %
1,10-phenanthroline	1	40
	2	4
1,7-phenanthroline	2	101
	8	95
4,7-phenanthroline	3	98
8 hydroxyquinoline-5-sulfonate	1	33
	5	15
EDTA	3	30
	5	2

[*] Similar results have been obtained on the enzymes from murine, simian, feline and RD-114 Type C RNA tumor viruses which have also been shown to be zinc enzymes (54).
[**] All rates measured at 25°.
[***] Enzyme activities in the absence, v_c, and in the presence of inhibitor, v.

See also references 101 and 102 for analogous studies.

Table VI

Analysis of Gel Exclusion Chromatography Fractions
(45 µl) Exhibiting Maximal Activity*

Fraction No.	Zn g at × 10^{11}	Protein µg	Zn/Protein g at/mole	Cu g at/mole	Fe g at/mole	Mn g at/mole
18	0.7	0.7	1.8	<.001	<.01	.06
19	1.5	1.4	1.9	<.001	<.01	.05
20	1.2	1.1	1.9	<.001	<.01	.05

*Calculations are based on a molecular weight of this polymerase of 1.8 × 10^5. Zinc and protein content are expressed as g atoms or µg per 5 µl aliquot, respectively. Analyses were performed in triplicate.

See also references 101 and 102 for analogous studies.

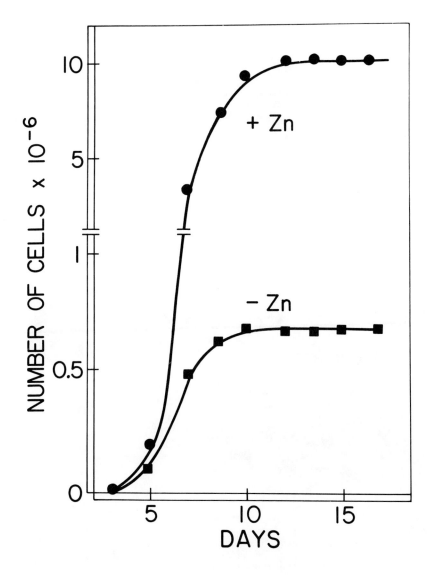

Fig. 1. Growth of zinc-sufficient (●) and zinc-deficient (■) E. gracilis grown in the dark. Zinc-sufficient medium contains 1×10^{-5} M Zn^{2+}; zinc-deficient medium contains 1×10^{-7} M Zn^{2+}. From (49).

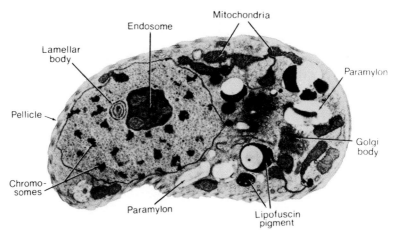

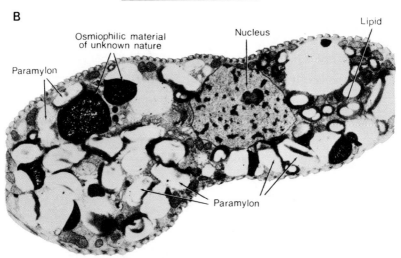

Fig. 2. Electron micrographs permitting comparison of the ultrastructure of Euglena grown 10 days with or without zinc, respectively. The zinc-deficient organism is significantly larger, its cytoplasm contains an abundance of paramylon and large masses of dense osmiophilic material presumably rich in lipid. The size difference is actually greater than it appears, since the micrograph of the zinc-deficient Euglena is shown at somewhat lower magnification so that the whole organism can be included. X 8500 and 6500, respectively.

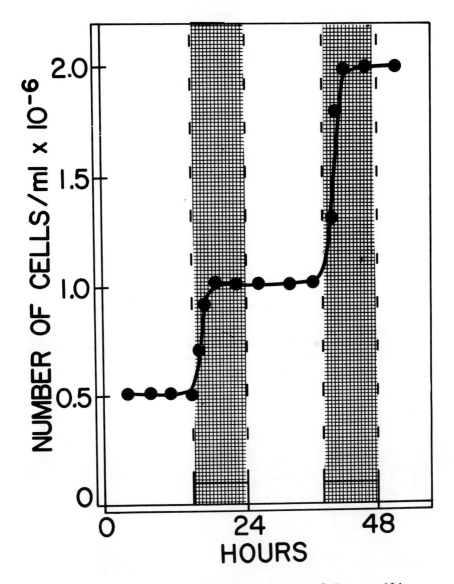

Fig. 3. Synchronously dividing culture of E. gracilis. The culture was exposed to alternating periods of light and dark, lasting 10 and 14 hours, respectively. Cells do not divide in the light periods. A burst of cell division occurs within 24 hours of entering the dark period. From (48).

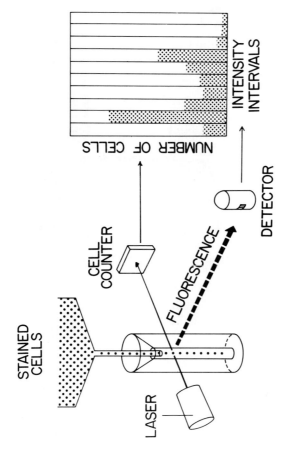

Fig. 4. Schematic diagram of Flow Cytofluorometer. From (48).

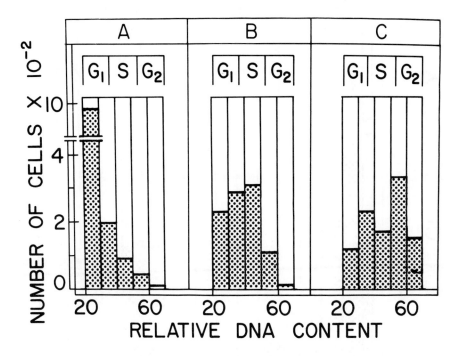

Fig. 5. Histogram of DNA content of synchronized cells harvested during the light period at 4, 8 and 14 hours. At 4 hours most cells are in G_1 and have not initiated DNA synthesis (Panel A). At 8 hours as the fraction of cells synthesizing DNA increases, the height of the G_1 peak decreases and that of the S phase increases (Panel B). At 14 hours, the majority of cells are in S or have reached G_2 and doubled their DNA content (Panel C). From (48).

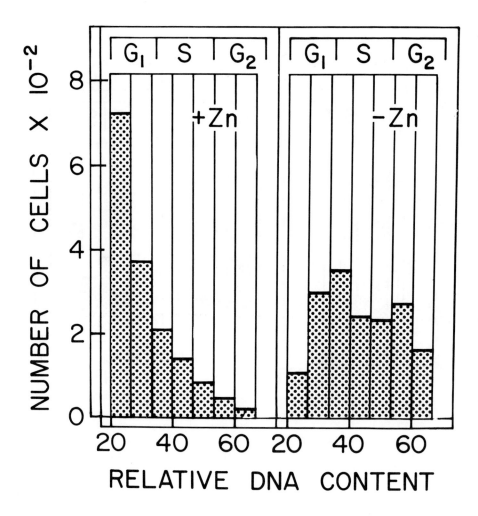

Fig. 6. Comparison of the histograms of DNA content of zinc deficient with stationary phase zinc sufficient E. gracilis. In stationary phase, the majority of sufficient cells are in G_1, with a small fraction in S. In contrast, nondividing deficient cells are mostly in S or G_2, with only a fraction in G_1. From (48).

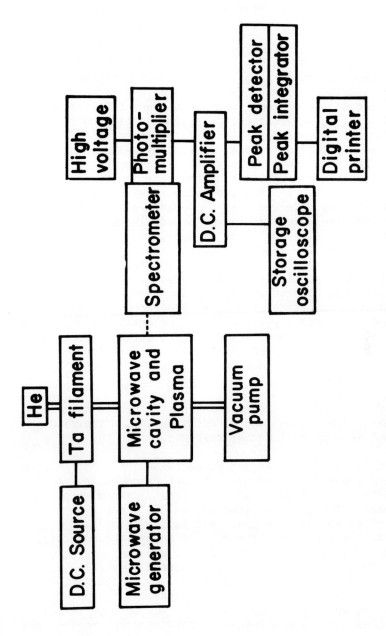

Fig. 7. Block diagram of microwave excitation spectrometer. From (72).

Discussion

Z. Brada, Papanicolaou Cancer Research Institute: I would like to comment on the possibility of using phenanthrolines in experiments in vivo . Dr. Mildvan mentioned yesterday some experiments demonstrating the effect of phenanthrolines on the growth of some tumors. o-Phenanthroline influences in vitro many enzymatic systems containing zinc or iron. During the years 1969 and 1970 we tried to influence the collagen formation in liver in vivo by o-phenanthroline treatment because it was known that this compound is an excellent inhibitor of proline hydroxylase in vitro. We were able to inhibit by this substance the fibrosis produced by ethionine. We observed that o-phenanthroline inhibited not only the development of fibrosis but also some other pathologic changes in the liver as well, including the formation of hepatoma. o-Phenanthroline itself has no toxic effect on normal rats at the same concentration. After feeding o-phenanthroline to normal rats for more than one year, we did not observe any pathologic changes and the growth of rats was stimulated. The frequency of spontaneous tumors was significantly decreased. The action of o-phenanthroline is not related to its chelating activity because the same effect can be produced by the non-chelator m-phenanthroline. Phenanthrolines influence also the activity of microsomal hydroxylases.

B. Vallee, Harvard Medical School: It would seem that your remarks are not related directly to this discussion and do not seem to call for my response.

A. Mildvan, Institute for Cancer Research: As I mentioned yesterday our effect of orthophenanthroine on the mouse tumor system that Yarbro studied is also obtained with metaphenanthroline, in confirmation of Dr. Brada's work, - so we rejected the view that this had anything to do with metal binding as mentioned. Phenanthroline molecules are planar, hence they can intercalate and can interact in many other ways and this is an important control that should be done in any biological study of orthophenanthroline. Hugh Creech in this regard made an interesting observation a number of years ago which may now be understandable. If one takes a two-arm nitrogen mustard, one can replace one of those arms with a phenanthroline type molecule and it is almost as effective as the two arm mustard. Now if the arm that has the phenanthroline has orthrophenanthroline it is about 10 times more effective as a mustard with a metaphenanthroline arm. But that is the only kind of experiment that I am aware of to implicate metal binding in such systems?

B. Vallee: In a way I regret that this discussion emphasizes chemotherapeutic considerations which I did not discuss. Actually, I did not intend to touch on that subject. I had intended to present the biological foundations of the problem; what their ultimate applications to normal and abnormal growth will be is anyone's guess at this time; but I would say that my major message is that there are whole areas of both biology and medicine that have grown up around the information relating to iron metabolism which Professor Theorell spoke about. Hematology and hemoglobin metabolism are based almost entirely on the chemistry of one metal, iron, and here zinc seemingly affects biology on a range as wide and far, and I would dare say, we might perhaps wish to consider that in evaluating future efforts.

S. Weinhouse, Temple University: What does zinc deficiency do to the zinc containing enzyme? Does it decrease the level of the apoprotein?

B. Vallee: A very good question. Very little if anything has been done on it. One study has examined the formation of alkaline phosphatase in zinc deficient systems. The interesting thing is that the apoprotein forms completely in a normal manner in that instance but is enzymatically inactive, of course. Thus, the existent genetic information results in the correct formation of the metal binding site, and its interaction with added zinc then results in immediate activity. I cannot think of another, similar experiment of this type at the moment. Incidentally, it is for that reason that I do not believe it is necessary to demonstrate reversible removal and restoration of zinc to a protein as proof that it is a metalloenzyme.

J. Schultz, Papanicolaou Cancer Research Institute: This is one of the most exciting papers that I have heard in a long time Dr. Vallee. However there are one or two things which I would like to have explained. You showed us a picture of a polymorphonuclear cell and told us that the zinc content was 0.35 something; on the other hand you showed us a lymphoblast with a value of .05. Now, I am not

a hematologist, but I am quite familiar with the fact that
the lymphoblast is not the same cell line as the polymorphonuclear cell. They are two different series and two different cells. That is one thing. The other thing is that the
lymphoblast had a .05 figure which you said was about 1/6 or
1/10 of the mature cell. But the lymphoblast is a fast
growing cell. The polymorphonuclear cell does neither
divide nor grow. It is mature. Here you have a cell rapidly
growing just like all cancer cells. It has only 1/10 the
amount of zinc that the mature cell has, which is not
growing. These facts are not in agreement with the stated
philosophy.

B. Vallee: This lecture could not address itself to
the details you have in mind, for which I apologize. The
present data are corrected, of course, for the mean corpuscular cell volume and hence, size. It is a fact that normal,
proliferating tissues are particularly prone to contain
large amounts of zinc. The values in zinc deficient cells
are grossly and significantly different.

M. Bade, Boston College: I realize that you were not
trying to be all inclusive in your listing of zinc enzymes,
but I might have a new one for you. Insects in molting have
to shed their old cuticle and it breaks down to a large
extent with the aid of a neutral metal chelator-sensitive
protease. If you feed orthophenanthroline to insects, then
they are unable to shed their cuticle so that they go through
a molt not being able to get out of the previous cuticle.
This is a well-known phenomenon from juvenile hormone experiments known as ecdysial failure. I would also like to
mention that we have verified the observations on how very
good orthophenanthroline is for rats. We thought that
orthophenanthroline would obviously be very bad and rats
obviously would not do well on it. Well, we fed it and they
grew 20% faster on the average and looked so healthy that
the controls looked somewhat sickly by comparison.

B. Vallee: I told you that zinc in good!

H. Petering, University of Cincinnati: Having worked
with metal chelating agents as chemotherapeutic agents for
some time it has become evident that these are not necessarily metal sequestering agents, but sometimes they are metal
delivering agents. They form complexes in vivo and then can
deliver metals to cells and tissue. This is the case with

respect to mono and dithiosemicarbazones. Thus, there is a different mechanism than the postulation of sequestering of essential metals, namely a metal and transport system.

B. *Vallee:* Thank you.

Z. *Chmielewicz, State University of New York:* Euglena is a vitamin B_{12}-requiring organism. There is a microorganism, Lactobacillus leichmanii, that also requires vitamin B_{12}. Deficiency of vitamin B_{12} produces morphological changes which resemble those which you show for Euglena. In one of your slides you listed various metals, metal content in zinc deficient and control Euglena. But I noticed that you did not list any cobalt. Have you looked at cobalt content of zinc deficient Euglena?

B. *Vallee:* Oh sure. It is well known, of course, that the growth of Euglena gracilis is vitamin B_{12} dependent, which has in fact served to assay this vitamin. However, the amount of cobalt present in vitamin B_{12} (4%) would not be detected by the techniques that we employed. The amount of cobalt in the total amount of vitamin B_{12} present would be far below the limit of its detection even by this sensitive method, so one cannot depend on such measurements. What I have said here for E. gracilis seems to hold true for M. smegmatis whose response to iron and zinc deficiency has also been studied, but not as extensively.

Z. *Brada, Papanicolaou Cancer Research Institute:* To the effect of orthophenanthroline in vitro, I would like to mention that Ciba Co. in Switzerland sells a preparation containing non-chelating derivatives of orthophenanthroline. This product controls intestinal infection. Apparently this effect is not caused by the chelating activity.

TARGET DIRECTED CANCER CHEMOTHERAPEUTICAL AGENTS

NATHAN O. KAPLAN
Department of Chemistry
University of California, San Diego

Abstract: We have been attempting to devise drugs based on their target specificity and which may have increased sensitivity for tumor cells. One such compound, prepared by others, is steptozotocin which appears to have some selectivity for islet cells, and has been used with varying success in treatment of insulinomas. This compound is a glucose derivative in which the glucose is linked to nitrosomethyl urea. The drug appears to bind to the glucose receptor in membranes and apparently the cytotoxic reagent is released into the cytoplasm. Evidence supporting this view will be summarized. Work on estradiol derivatives of this type will also be described.

Binding of a number of hormones covalently to insoluble matrices has been carried out by B. Venter and J.C. Venter in our laboratory. The procedures used for this binding will be presented. These immobilized entities have been shown to react specifically with cultured tumor cells. The importance of the type of linkage in the interaction of cells with the matrix will be discussed. Biochemical changes resulting from the interaction of cells with the matrix have been quite dramatic.

The structural-functional relationships of various toxins will be surveyed. A number of analogues of diphtheria toxin have been prepared by Dr. J. Everse of this laboratory and their properties tested. The possible use of toxins in cancer chemotherapy will be discussed. Experiments using human tumors growing in athymic mice also will be evaluated.

Finally a discussion will be given of peptides as potential vehicles for delivery of drugs.

INTRODUCTION

What I would like to discuss today are studies pertaining to the possibility of a group of compounds selectively and specifically directed which may have a greater cytotoxic effect on malignant tumor cells as compared to normal cells. During the past twenty-five years many agents have been developed which show preferential cytotoxicity for tumor cells and have been used with varying success in cancer chemotherapy. Although the results have certainly not been spectacular, they do show that a number of agents can exhibit growth of cancer cells under conditions where normal cells are relatively unharmed.

Most of the drugs used are antimetabolites or antibiotics which directly or indirectly inhibit protein or nucleic acid synthesis. For example, we know that a drug such as methotrexate can act as a potent anti-tumor agent without inducing an overall toxicity on patients or experimental animals. This point should be emphasized because of the recent use of (toxic) levels of the antifolate followed by the introduction of the citrovorum factor (1). This type of rescue therapy now being used in patients was first studied a number of years ago in experimental animals by Goldin and associates (2).

In discussions with Professor Morris Friedkin, he has pointed out to me that as far as one can ascertain there is no molecular enzymatic basis for the rescue phenomenon to selectively protect the normal cells and not the malignant cells. Recently, Halpern and associates (3) have advanced the hypothesis that tumor cells lack the enzyme that converts homocysteine in the presence of methyl tetrahydrofolate to methionine and this may bring about the difference between normal and malignant cells because the lack would lead to a deficiency of tetrahydrofolate required for other essential reactions such as thymidylic acid synthesis.

As an enzymologist, I had high hopes twenty-five years ago that one might be able to detect differences in the enzyme patterns which would allow for the rational use of drugs in the treatment of cancer. When isoenzymes were found, many of us thought that because of the difference in catalytic properties that selective inhibition of tumor cells could be accomplished. Although I believe some very interesting information has been obtained from such studies, the results in general have not been as rewarding as one might have expected. The limited success of some of the compounds now being used may be somewhat related to the concentration of enzymes or isoenzymes and not to some distinct patterns which might be characteristic of tumors. It is true that a number of malignant cells cannot synthesize certain non-essential amino acids such as glutamine (4) or serine (5), but it also appears that a number of "normal" cells can also not achieve this synthesis. Furthermore, this problem becomes complicated because of the presence of these amino acids in serum where they may be initiated from synthesis in other tissues as well as being components in the diet. What I am trying to say here is that a straight enzymological approach has not been of practical use even though many promising trends and significant basic information has been obtained by the multitude of studies which have been carried out during the past quarter of a century.

NICOTINAMIDE ANTAGONISTS

The problem of tissue specificity first interested us during the course of our studies with NAD analogues. These analogues were synthesized by the exchange reaction catalyzed by mammalian NADases, which is a glycohydrolase acting hydrolytically or promotes exchanges with other pyridine bases. The hydrolysis reaction is inhibited by nicotinamide through this exchange mechanism which is summarized in Fig. 1 (6). The exchange reaction involves the transfer of the

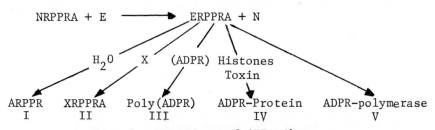

Fig. 1. Reactions of ADP-ribose

ADPR moiety to the pyridine acceptor. This type of reaction has received considerable attention during the past few years. There is an enzyme present in the nucleus which promotes the formation of poly ADPR where NAD is the substrate. The polymer appears to be linked to histones or other nuclear proteins and a great deal of work is now underway to show its role in regulation of DNA synthesis (7). Diphtheria toxin can catalyze the adenosinediphosphate ribosylation of translocase and thereby inhibit protein synthesis (8) (some aspects of the toxin activity will be described later in this paper). Adenosinediphosphate ribosylation of the RNA polymerase of E. coli has recently been found to be one of the early events occurring after phage infection (9). The donor molecule is again NAD^+ and the modification of the polymerase has been implicated in the change of this enzyme's specificity. Very recently NAD^+ has been found to be essential in the activation of adenylcyclase by cholera toxin, although as yet no definite evidence has been found for adenosinediphosphate ribosylation of the membrane (10). It is evident from the above discussion that NAD is not only of importance in oxido-reduction but may play a significant role in regulation through the transglycosidase reaction.

I would like to discuss two nicotinamide antagonists as examples of problems of specificities which one encounters in attempting to develop a useful cancer chemotherapeutic agent. These are 6-amino nicotinamide and 3-acetyl pyridine (11). Their structures are given in Fig. 2. Both compounds form

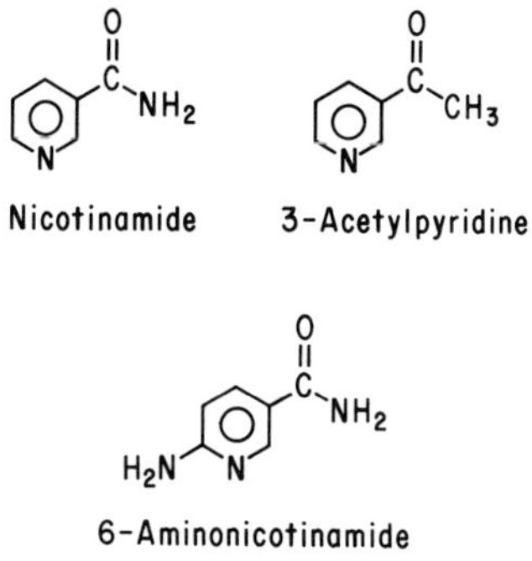

Fig. 2. Structures of nicotinamide and antagonists.

NAD analogues through the exchange reactions. Tables 1 and 2 illustrate that when these compounds are injected into mice

TABLE 1

NAD and acetylpyridine NAD levels of L1210 tumor

Animal number	Controls (saline injected) NAD	Animal number	Acetylpyridine injected NAD	Acetylpyridine injected acetylpyridine
1	155	7	67	60
2	142	8	102	53
3	131	9	55	72
4	145	10	47	69
5	136	11	73	48
6	151	12	64	81
Average	143		68	64

TABLE 2

Effect of 6-amino nicotinamide on NAD levels of mouse L1210 cells

	NAD	6-amino nicotinamide NAD
Saline injected	162 ± 30 (10)	0
6-amino nicotinamide		
25 mg/kg	125 ± 18 (8)	30 ± 11 (8)
60 mg/kg	105 ± 23 (11)	53 ± 12 (9)

carrying L1210 tumors that the coenzyme analogues are formed in the tumor *in vivo*. There is a decrease in NAD^+ concentration which correlates roughly to the amount of analogue produced. It should be noted that much higher levels of 3-acetylpyridine were used in these experiments since it is known that the 6-amino nicotinamide is considerably more toxic. This difference in toxicity may be related to the fact that 3-acetylpyridine can be converted to nicotinamide in the liver.

Both compounds appear to extend the survival time of the L1210 mice as indicated in Table 3. The tumors are greatly

TABLE 3

Protection of L1210 mice by injection of
3-acetylpyridine or 6-amino nicotinamide

	Mean survival (days)	Number of animals
Saline controls	11.2	15
250 mg 3-acetylpyridine[a]	17.5	15
25 mg 6-amino nicotinamide	16.1	10

[a] Injected daily after introduction of tumor cells.

decreased in size when compared to control animals. Administration of compounds over a number of days appears to produce toxicity which is indicated by neurological changes. We were never able to find a dose of the antimetabolites which would completely eradicate the tumor without producing toxicity.
It should be pointed out that administration of nicotinamide with either of the compounds abolishes their effects on survival, or their inhibition of tumor growth. Perhaps the problem should be reinvestigated in light of the results with the methotrexate rescue by administration of the citrovorum factor.

The symptoms of toxicity with either 3-acetylpyridine or 6-amino nicotinamide are usually dissociated with dramatic neurological changes. Hicks (12) in studies carried out a number of years ago found that injection of 3-acetylpyridine into rats or mice induced leisons of the hypothalmic area but only minor changes in other areas of the brain. Marked pathological alterations were found primarily in the supraoptical nucleus and the hippocampus (13). Nicotinamide administration protects completely against the brain pathology resulting from 3-acetylpyridine injection. In the studies of Coggeshall and MacLean (13) the leisons were almost exclusively limited to areas CA3 and CA4 of the hippocampus. These authors suggested that 3-acetylpyridine might be used as a probe for studying behavior related to the various areas of the hippocampus. It is of interest to note that the brain pathology induced by 3-acetylpyridine does not occur when this compound is injected into fetal or newborn animals.

This observation may be related to the findings of Burton (14) that the NADase activity of the brain is present only after birth.

In contrast to the changes produced by 3-acetylpyridine, administration of 6-amino nicotinamide resulted in relatively little or no damage to the hypothalamus but profound changes were found in the cerebellum and spinal cord (15). Hence, if you want to destroy the hypothalmic region you use 3-acetylpyridine. If you want to destroy the hind brain, 6-amino nicotinamide would be the agent of choice. It is important to stress that the pathology induced by both compounds can be antagonized by nicotinamide and both can be converted to analogues of NAD.

The above results were most interesting in the sense that different types of nerve cells varied in their response to two different antagonists of nicotinamide. Why the difference? In our first attempt to obtain information on this problem, we were unable to detect either of the coenzyme analogues in whole brain or in different areas of brain by chemical methods. But more recently using labelled 3-acetylpyridine we were able to demonstrate some analogue formation in the hypothalamus area of the rat as well as in the cerebellum. Good quantitative data were difficult to obtain because of the relatively low levels of incorporation. The results, however, suggested that there was no striking differences in levels of the coenzyme analogues as compared to NAD levels (Table 4). There also appears to be no significant

TABLE 4

Effect of nicotinamide on pyridine nucleotide levels of the brain

	Hypothalmus		Cerebellum	
	gamma/gram fresh weight			
	NAD	analogue	NAD	analogue
3-acetylpyridine[a]	186	21	114	14
6-amino nicotinamide[b]	192	12	130	10

[a] Injected intraperitoneally 500 mg/Kg body weight.

[b] Injected intraperitoneally 50 mg/Kg body weight.

difference in the two areas of the brain when 6-amino nicotinamide is injected intraperitoneally. What these results suggest is that the two areas of the brain react differently to the two coenzyme analogues, if indeed the analogues are responsible for the pathology.

The findings described above show some relationship to the classical teratogenic studies of Landauer (16), who observed upon injection of 3-acetylpyridine into the yolk sac of 96 hour old chick embryos, they produced skeletal muscle hypoplamic so that on hatching little or no leg musculature was present. In contrast when 6-amino nicotinamide was injected in the yolk, no abnormalities of the muscle system were observed, but striking malformations of the bone and cartilage were found in the hatched embryos. The nicotinamide antagonists were producing different teratogenic effects: The teratogenicity of both compounds was completely abolished if they were coinjected with nicotinamide.

In collaboration with Dr. Arnold Caplan and the late Professor Edgar Zwilling at Brandeis, we attempted to give a molecular explanation for the difference in abnormalities resulting from the presence of the two analogues (17). Caplan established cultures of developing embryonic chicken muscle as well as chondrogenic cells. It was found that 3-acetylpyridine was lethal to the myoblasts in culture as might be predicted by Landauer's work on the whole embryo. In particular, proliferation was inhibited: However, cells affected by 3-acetylpyridine still retained the potential to fuse and form multinucleated cells, which is a normal event in muscle development. The chondrogenic cells, however, were not inhibited, but on the contrary their proliferation was greatly enhanced by the presence of the same concentration of 3-acetylpyridine which produced the degenerative changes in the muscle cultures. Both the potential effect on the cartilage cells and the degeneration of the myoblasts were reversed if nicotinamide was added simultaneously. The results of this study are schematically given in Fig. 3.

Since both cartilage producing cells and the myoblasts were from mesodermal cells, it is possible that NAD concentration may effect the differentiation into the two types of cells. When the NAD concentration is higher, the myoblasts are formed and stimulated; a low concentration of NAD would favor the proliferation of cartilage. It may be reasonable to assume that if NAD levels are influenced by 3-acetylpyridine, then normal development of the muscles could be

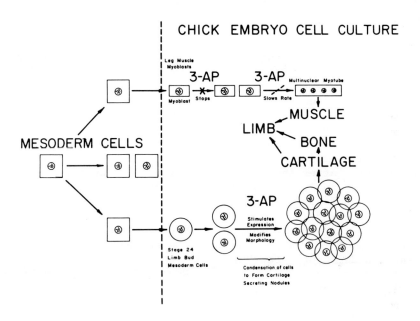

Fig. 3. Schematic representation of 3 acetylpyridine (3-AP) on limb development.

controlled by the coenzyme level which would result in sustaining muscle formation while at the same time inhibiting extensive cartilage formation. More recently Caplan and his associates have described this action as the result of the formation of poly ADPR complexes (18).

Studies such as described above indicate that regulatory mechanisms in different cells vary and that an understanding of the many phases of regulation in malignant cells as compared to normal cells may lead to drugs which will be beneficial in chemotherapy.

Before leaving the subject of the pyridine nucleotides, I would like to summarize studies which were carried out in collaboration with Dr. David Gardner and Professor Gordon Sato on the effects of 3-acetylpyridine on cells grown in culture (19). Control cultures grow normally when there are equimolar concentrations of 3-acetylpyridine and nicotinamide. In order to show the effects of nicotinamide deficiency, a second subculture of the vitamin deficient medium is essential. After the second subculture, the deficiency leads not only to an inhibition of growth, but a marked decrease in pyridine nucleotide levels. Steroid production is reduced by approximately 50% in the nicotinamide deficient cells. It was found that 3-acetylpyridine induced lethality was inversely proportional to the pyridine nucleotide levels. Hence, a decreased coenzyme level leads to a strong cytotoxic action of the 3-acetylpyridine. Reconstitution of nicotinamide in the depleted media causes a rapid increase in the NAD^+ concentration and complete protection against the toxic effects of 3-acetylpyridine. It is of interest to note that although high levels of nicotinamide will induce a marked elevation of NAD^+, toxic effects are observed with the higher concentration of vitamin. Furthermore, the reduced capacity to synthesize steroids was not restored on addition of nicotinamide even though the NAD^+ concentration and cell division rates were normal. Somehow the lowered NAD^+ induced a permanent change in the steroid synthesis potential. It is not known as yet whether these permanent changes are due to selection or mutation.

More recently, we have found similar effects with HeLa cells; the toxicity of the antagonist being directly correlated to the NAD levels (Table 5). What these experiments imply is that the cytotoxic effects of the antimetabolites are related to the concentration of metabolite in a particular cell. Hence one might expect that cells with low levels of NAD will be more sensitive to antagonists of nicotinamide. The HeLa cells appear to be more nicotinamide dependent than are the adrenal tumor cells.

TABLE 5

Relationship of NAD^+ concentration to 3-acetylpyridine toxicity in HeLa cells

NAD^+ µg/mg protein	Inhibition of growth 3×10^{-5} M 3-acetylpyridine percentage
2.63	<5
1.78	38
1.14	62
0.82	69
0.31	>95

STREPTOZOTOCIN

One of the interesting compounds which has been used in cancer chemotherapy is the antibiotic streptozotocin. The structure of this compound is given in Fig. 4. The

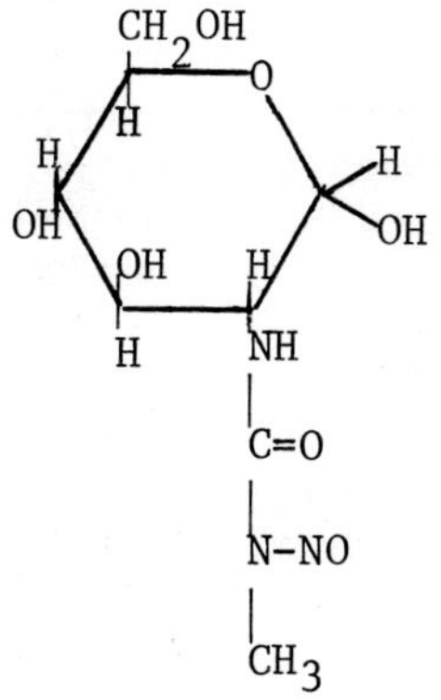

Fig. 4. Structure of streptozotocin.

antibiotic is essentially a glucose derivative of the cytotoxic agent nitrosomethyl urea. Studies with this compound show that it has anti-tumor action. However, it was found that the injection of this drug into experimental animals produced a permanent diabetes (20). One of the interesting features is that the antibiotic lowers the pyridine nucleotide levels both in liver as well as in the islet cells (21, 22). Preinjection of nicotinamide will prevent the decrease of NAD as well as preventing the occurrence in the diabetic state. Nicotinamide will not protect against the diabetes induced by alloxan if the vitamin is given within a certain time range. Although nicotinamide may protect against streptozotocin with respect to the pyridine coenzyme levels, the compound still has a cytotoxic effect on certain tumors in vivo in the presence of the vitamin. This phenomenon appears to be best explained by lowered drug toxicity, but not therapeutic activity. The exact reasons for this selectivity are as yet not clear.

Streptozotocin has been used with varying success in the treatment of human insulinomas; apparently the antibiotic is most effective on this type of tumor (23). The possible differential exerted on the beta cells may be related to the higher sensitivity of the glucose receptor in pancreas cells as compared to other cells. Nitrosomethyl urea does not have this selective action on the islets. Hence because of the glucose moiety, the cytotoxic action of the alkylating agent becomes more specific.

In preliminary experiments we have observed that the streptozotocin can be bound to the glucose receptor in the islet cells. However, using doubly labeled streptozotocin we have not been able to demonstrate unequivocally that there is a cleavage of the drug once it is attached to the receptor. We believe that a detailed study of the action of streptozotocin may lead to new approaches on development of drugs which are specific for particular types of malignant cells.

Barbara Venter in our laboratory has synthesized the alkeran derivative of estradiol. This compound which appears to bind to the estrogen receptor is now under study as to whether it has selective toxicity on mammary tumors. The work has not reached a stage as yet where any conclusion can be made.

IMMOBILIZED DRUGS

I would like now to turn to some of our recent studies on immobilized drugs which may have use as models for the development of target specific chemotherapeutic drugs.

For a few years we have been investigating the biological effects of various catecholamines and their derivatives immobilized to glass largely through the diazonium linkage (24-27). The structure of such derivatives is given in Fig. 5. The length of the spacer does not appear to be a factor

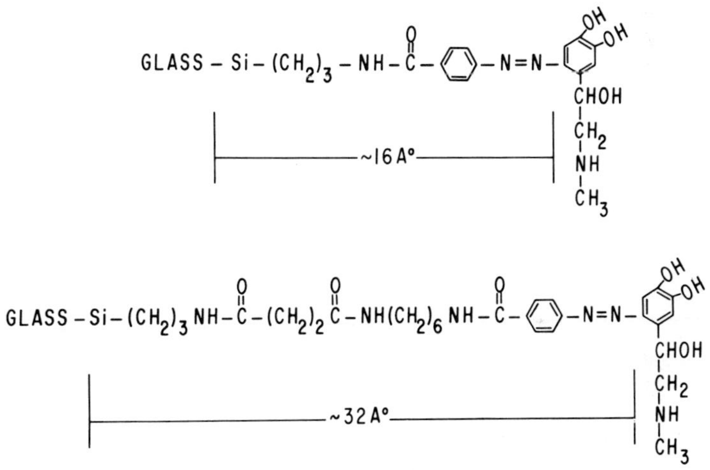

Fig. 5. Structure of glass bound catecholamines.

on the biological activities we have measured. Although it does appear that chain lengths of at least 5 and 6 are essential for activity. In general, the glass-catecholamines have a much longer period of action than do the corresponding free compounds. This is illustrated in Fig. 6 when the immobilized catecholamine is directly applied to the pacemaker area

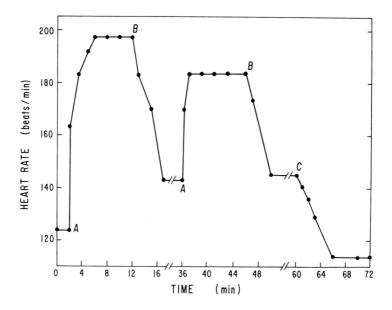

Fig. 6. An example of the influence of "isoproterenol- and propranolol-on-glass" on heart rates in the dog. (A) denotes the point at which "isoproterenol-on-glass" was applied to the sinoatrial node of the dog heart; (B) represents points at which the glass beads were removed from the heart; (C) denotes the point at which "propranolol-on-glass" was applied to the sino-atrial node of the heart.

of the dog heart (24). It is of interest to note that the antagonist propranolol when bound to the glass can induce a prolonged decrease in the heart rate (27). In Fig. 7, the effects of immobilized and soluble isoproterenol on cat papillary muscle are compared. It is of interest that the increased force induced by the free isoproterenol lasts roughly only about 30 minutes whereas the increase resulting from the immobilized drug lasts at least 6 hours and we have observed preparations where the elevation in force has been retained for a period of 10 hours.

It is of importance that the effects of the catecholamine-glass occur only when the beads are physically in contact with the muscle. We have ruled out the possibility that leakage of free catecholamine from the beads is the cause of

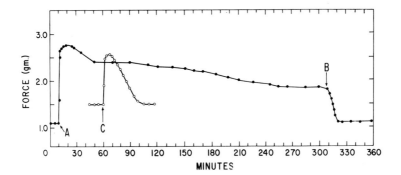

Fig. 7. The effect of the addition of immobilized isoproterenol (A) to a cat papillary muscle is compared with that of addition of 1 µM isoproterenol solution (C) to the bath of a separate muscle. (B) Denotes the time when the isoproterenol beads were removed from the muscle.

the activity (27). However, we have not ruled out the possibility that high local concentrations of free drug are responsible for the effect. Our studies indicate that the free native drug does not appear in any substantial amounts. There is the possibility that certain derivatives which result from the glass dissolving may be the active species. At any rate we cannot say with any degree of certainty at this time that the immobilized drug is acting while still attached in a covalent linkage to the glass (27).

Dr. Craig Venter has found that although the immobilized catecholamine will give roughly the same increase in contraction as does the free compound, cyclic-AMP is not increased when bead preparations are used under the same conditions where the free drug produces a significant elevation in cyclic nucleotide (26).

The many experiments which we have carried out with the immobilized catecholamine indicate that a compound bound to a receptor may act for long periods of time when it is not inactivated or diffused away from the receptor. Hence the glass bead isoproterenol or epinephrine will induce large increases in the cyclic-AMP of glioma cells but will only produce small changes on other lines of cells which have low concentrations of the catecholamine receptors.

In an effort to resolve whether the catecholamine can still be active when covalently bound, we have turned to the study of soluble synthetic peptides in collaboration with Professor Murray Goodman and Drs. Michael Verlander and Craig Venter (28). These compounds are mixed copolymers of amino phenylamine and a derivative of glutaric acid. The isoproterenol was linked by a diazonium bond. The polymers were highly purified and several different molecular weight entities were isolated ($\sim$1,500, 3,000, 6,000, 10,000). All sizes of the drug-copolymer preparation were found to be active, and all the evidence indicated that the polymers were active when the drug was covalently bound when tested on isolated cat papillary muscle or perfused guinea pig hearts (28). These results strongly suggest that at least some drugs have activity while linked to a polymer.

It is of interest to note that in recent experiments on the intact dog that increases in the heart rate were observed with isoproterenol polymers (29). For example the smaller the polymer, the longer was the extension of increased heart rate. Free isoproterenol has a half life of about 3 minutes; the 110,00 M.W. polymer has a half life of 5 minutes, the 3,000 about 14 minutes and the small molecular weight entity (1,500) about 22 minutes (29). Hence the size of the peptide appears to be of critical importance in extending the activity of the drug. These preliminary experiments have been very encouraging and Professor Goodman's and our group are now deeply involved in an effort to synthesize and study polymeric drugs which may be useful for basic studies as well as for clinical use.

An interesting use of immobilized drugs and hormones has been for probing receptor sites on intact whole cells. In the case of the bound catecholamines, it was found that rat glioma cells (C6) would strongly adhere to either isoproterenol glass beads or Sepharose (Fig. 8), whereas attachment to

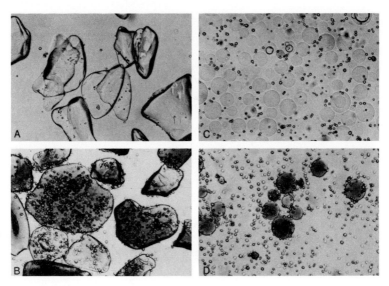

Fig. 8. Affinity isolation of cultured glial tumor cells utilizing isoproterenol covalently coupled to glass and sepharose beads. Photomicrographs in Panel (A) show control arylamine glass beads following a 15-min incubation with 10^6 cells/ml. (B) shows the results of an identical incubation with isoproterenol-glass beads. (C) shows control sepharose arylamine beads under identical incubation conditions as in (A). (D) shows glial cell binding to isoproterenol-sepharose beads after a 15-min incubation.

other mammalian cells such as the rat adrenal tumor cells (Y-1), HeLa cells, or rat pituitary tumor cells (GH_3) (30). Binding was prevented by incubation of the cells with concentrations of propranolol although higher concentrations of the drug were required for promoting inhibition of the attachment of the cells to the isoproterenol glass preparation. Norepinephrine linked through the catecholamine sidechain did not bind cells, thereby demonstrating certain specific structural requirements for drug-cell interactions.

The rat pituitary cells have been found to bind to immobilized triiodothyronine with greater efficiency than the glioma or adrenal tumor cells. On the other hand, the immobilized ACTH binds the adrenal cells more tightly and thereby indicates strong evidence for specific drug and hormone receptors on the plasma membrane.

DIPHTHERIA TOXIN

Because of our studies with drug carriers, we have in the past year and a half developed an interest in diphtheria and related toxins. We have been studying for some time the mechanism by which the toxin catalyzes the transfer of ADPR to the translocase (EL_2) which causes the inactivation of this enzyme according to the following equation:

$$NRPPRA + EL_2 \xrightarrow{toxin} EL_2RPPRA + N$$

The whole toxin is a protein which can be nicked to give a molecule with two polypeptide chains held together by a single disulfide bond (8) (Fig. 9). This insulin-like molecule

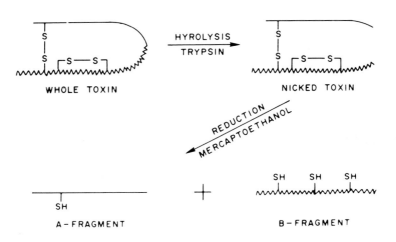

Fig. 9. Diphtheria toxin structures.

can be reduced to give individual polypeptide chains A and B. The A chain has been established to be the factor which produces the inhibition of protein synthesis. The A moiety is now being sequenced by Dr. Collier of UCLA. We have recently been able to separate the A and B chains in relatively large quantities with both chains remaining. We are now studying the properties of the B chain. The A peptide is inactive on whole cells and is not toxic to animals. Since the A chain acts in broken cells, it is apparent that the disulfide linkage is essential for transport of the toxic moiety into the cells. We are now exploring the possibility of using the B chain as a vehicle for transporting substances into cells.

Barbara Venter and Craig Venter in our laboratory have shown that HeLa cells are bound tightly to diphtheria toxin coupled to Sepharose beads via an amide bond (30). This binding is inhibited by prior incubation of the Sepharose toxin with purified antitoxin. When the toxin was bound through an azo linkage to the Sepharose, the cells did not bind.

We have now found that when unnicked toxin is used, there is a much stronger affinity of the cells for the Sepharose derivative than when the nicked toxin is immobilized. As yet we have found no binding of the cells when the A or B fragments are linked either by themselves or together. These results suggest that the affinity for the cell by the toxin is inherent in the disulfide bond. Details of these studies will be published elsewhere (31).

It has been reported (32-34) that diphtheria toxin can be used in the treatment of mouse tumors (in particular ehrlic ascites cells) and that complete regression of the tumors could be sometimes observed.

In fact Iglewski and Riggenberg (32) reported a ten thousand-fold greater sensitivity to the diphtheria toxin of tumor cells as compared to normal cells when measured by inhibition of protein synthesis in whole cell preparations. This greater sensitivity of neoplastic cells occurred both in mouse as well as human cell lines.

Abrin and ricin are toxins which have structures closely related to diphtheria toxin and a disulfide bridge is essential for animal toxicity and in inhibiting protein synthesis in whole cells. The structure of these toxins is schematically given in Fig. 10 (35).

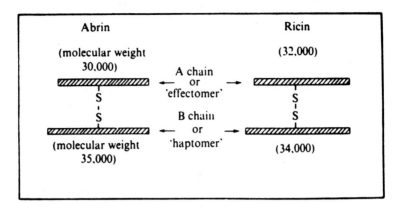

Fig. 10. Structure of abrin and ricin [Olnes et al. (35)].

The A fragment of both toxins or as Olnes et al. have described as the effectomer molecule is very inhibitory to protein synthesis when all extracts are used but without effect on whole cells or toxicity to animals. Hence A and B fragments of the three toxins appear to be related in their mechanism of action.

In fact Lin et al. (36) have shown that both abrin and ricin can cause complete regression of Ehrlich Ascites tumor in the mouse. This group of investigators (37,38) have also reported successful treatment of human cancer patients by use of either ricin or abrin.

The above studies stimulated us to prepare derivatives of the toxins which might be less toxic and more selective on cancer cells. During the course of this, a recent article by Pappenheimer and Randall (39) appeared stating that diphtheria toxin was unlikely to be an effective anti-cancer agent. These authors claimed that they were not able to repeat Iglewski's and Rittenberg's findings. They were unable to show any inhibition on protein synthesis on Ehrlic cells in a tissue culture medium, although they found some regression of the Ehrlic ascites tumor in the mouse. When the toxin was administered, we thought it of importance for us to ascertain whether Iglewski and Rittenberg (32) or Randall and Pappenheimer were correct (39). Although at first we were unable to show any inhibition of protein synthesis of the Ehrlich cells in a tissue culture medium, by changing conditions and using the appropriate number of cells, Janice Beglau of our laboratory was able to demonstrate that inhibition of protein synthesis was relatively the same for the Ehrlich cells when compared to HeLa cells although the time for inhibition to set in was slightly longer with Ehrlic cells. It should be pointed out that man is considerably more sensitive to the diphtheria toxin than are the rodents.

Mr. Stolzenbach and Dr. Everse have shown dramatic regression of the Ehrlic tumors with a certain number of "cures" both in normal Swiss mice as well as in athymic mice. Dose level and scheduling appear to be like other cancer chemotherapeutic agents, important factors in causing dramatic effects on the tumor.

Dr. Vincenzo Buonassisi and Patricia Viancour have demonstrated that injection of a single dose of toxin can cause complete regression of human tumor growing in the athymic mouse. Different human tumors appear to have different sensitivities to the diphtheria toxin. These preliminary studies although done in the mouse suggest that detailed investigation could lead to diphtheria toxin or its derivatives becoming effective useful anti-cancer agents in man.

The work with the toxins is a somewhat different approach towards developing cancer chemotherapeutic drugs. The toxin action is dependent on its ability to interact with the plasma membrane (40). It has become increasingly apparent that cell surface differences do exist between normal and neoplastic cells. The toxins are examples of how one might exploit the differences in membranes. Certainly other types of molecules could be tailored for cytotoxic activity by making use of the knowledge of the differences in membranes.

It is my belief that as we further understand membrane structure and function, drugs will be developed which are more target directed for cancer cells. It is also my opinion that emphasis on chemotherapy should be focused in this direction and our laboratory is now applying efforts towards this goal.

ACKNOWLEDGEMENT

Parts of this work were supported by grants from the American Cancer Society (BC-60-Q) and the National Institutes of Health (USPHS CA 11683).

REFERENCES

(1) N. Jaffe, M.B.B. Ch, E. Frei III, D. Traggis and Y. Bishop, N.E. J. Medicine 291 (1974) 994.

(2) A. Goldin, N. Mantel and S.W. Greenhouse, Cancer Res. 14 (1954) 43.

(3) R.M. Halpern, B.C. Halpern, B.R. Clark, H. Ashe, D.M. Hardy, P.J. Jenkinson, S.C. Chou and R.A. Smith, Proc. Nat. Acad. Sci. U.S.A. 72 (1975) 4018.

(4) R.E. Neuman and T.A. McCoy, Science 124 (1956) 124; E.E. Haley, G.A. Fisher and A.D. Fisher, Cancer Res. 21 (1961) 532.

(5) J.D. Regan, H. Vodopick, T. Susuma, W.H. Lee and F.M. Vaulcon, Science 163 (1969) 1452.

(6) N.O. Kaplan in Poly(ADP-Ribose), ed. M. Harris (Fogarty International Proc. #29, 1975) p. 5.

(7) T. Sugimura in Progress in Nucleic Acid Research and Molecular Biology, Vol. 13, eds. J.N. Davidson and W.E. Cohn (Academic Press, New York, 1973) p. 127; M. Smulson, Science 188 (1975) 1223.

(8) A.M. Pappenheimer and D.M. Gill, Science 182 (1973) 353.

(9) C.G. Goff, J. Biol. Chem. 249 (1974) 6181; H. Rohrer, W. Zillig and R. Mallhammer, Eur. J. Biochem. 60 (1975) 227.

(10) D.M. Gill, Proc. Nat. Acad. Sci. U.S.A. 72 (1975) 2064.

(11) N.O. Kaplan in The Neurochemistry of Nucleotides and Amino Acids, eds. R. Barsky and D.B. Towers (Wiley, New York, 1960) p. 70.

(12) S.P. Hicks, Amer. J. Pathol. 31 (1955) 189.

(13) R.E. Coggeshall and P.D. MacLean, Proc. Soc. Exp. Biol. Med. 98 (1958) 687.

(14) R.M. Burton, J. Neurochem. 2 (1957) 15.

(15) S.S. Sternberg and F.S. Phillips, Science 127 (1958) 644.

(16) W. Landauer, J. Exp. Zool. 136 (1957) 507; W.E. Landauer and E.M. Clark, J. Exp. Zool. 151 (1962) 253.

(17) A.I. Caplan, E. Zwilling and N.O. Kaplan, Science 160 (1968) 1009.

(18) A.I. Caplan and R.J. Rosenberg, Proc. Nat. Acad. Sci. U.S.A. 72 (1975) 1852.

(19) D.A. Gardner, G.H. Sato and N.O. Kaplan, Develop. Biol. 28 (1972) 84.

(20) P.S. Schein, D.A. Cooney and M. Vernon, Cancer Res. 27 (1967) 2324.

(21) P.S. Schein and S. Loftus, Cancer Res. 28 (1968) 1501.

(22) C.K. Ho and S.A. Hashim, Diabetes 21 (1972) 789.

(23) L.E. Broder and S.K. Carter, Ann. Internal. Med. 79 (1973) 101; L.E. Broder and S.K. Carter, Ann. Internal. Med. 79 (1973) 108.

(24) J.C. Venter, J.E. Dixon, P.R. Maroko and N.O. Kaplan, Proc. Nat. Acad. Sci. U.S.A. 69 (1972) 1141.

(25) J.C. Venter, J. Ross, Jr., J.E. Dixon, S.E. Mayer and N.O. Kaplan, Proc. Nat. Acad. Sci. U.S.A. 70 (1973) 1214.

(26) J.C. Venter, J. Ross, Jr. and N.O. Kaplan, Proc. Nat. Acad. Sci. U.S.A. 72 (1975) 824.

(27) N.O. Kaplan and J.C. Venter in Proc. Sixth Internal. Cong. of Pharm., Vol. 5, eds. J. Tuomisto and M.K. Paasonen (Helsinki, 1975) p. 73.

(28) M.S. Verlander, J.C. Venter, M. Goodman, N.O. Kaplan and B. Saks, Proc. Nat. Acad. Sci. U.S.A. (April, 1976) in press.

(29) J.C. Venter, S. Sasayama, J. Ross, Jr., M.S. Verlander, M. Goodman and N.O. Kaplan (1976) in preparation.

(30) B.R. Venter, J.C. Venter and N.O. Kaplan, Proc. Nat. Acad. Sci. U.S.A. (1976) in press.

(31) F.E. Stolzenbach and N.O. Kaplan (1976) in preparation.

(32) B.H. Iglewski and M.B. Rittenberg, Proc. Nat. Acad. Sci. U.S.A. 71 (1974) 2707.

(33) S. Buzzi and L. Buzzi, Cancer Res. 33 (1973) 2349.

(34) S. Buzzi and L. Buzzi, Cancer Res. 34 (1974) 3481.

(35) S. Olnes, K. Refsnes and A. Pihl, Nature 249 (1974) 627.

(36) J.-Y. Lin, K.-Y. Tseng, C.-C. Chen, L.-T. Lin and T.-C. Tung, Nature 227 (1970) 292.

(37) T.-C. Tung, C.-T. Hsu and J.-Y. Lin, J. Formosan Med. Assn. 70 (1971) 569.

(38) C.-T. Hsu, J.-Y. Lin and T.-C. Tung, J. Formosan Med. Assn. 73 (1974) 526.

(39) A.M. Pappenheimer and V. Randall, Proc. Nat. Acad. Sci. U.S.A. 72 (1975) 3149.

(40) G.L. Nicolson, M. Lacorbiere and T.M. Hunter, Cancer Res. 35 (1975) 144.

Discussion

R. Leif, Papanicolaou Cancer Research Institute: The fact that alkeran derivative of estradiol binds to estrogen binding sites and is a dye and perhaps even fluorescent, provides instant implications in diagnosing the estrogen sensitivity of breast and other tumors. Estrone, if I remember rightly is also a fluorescent compound. Dr. B. Watson and I probably will be exploring the use of estrone as a fluorochrome soon. Have you published anything on the binding of alkeran estradiol yet?

N. Kaplan, University of California: No. Barbara Venter is completing this work for her thesis and she soon will be publishing the results.

R. Leif: Do you know if this compound fluoresces - or any of the other estrogen derivatives which bind? Has she tried any other derivatives that bind?

N. Kaplan: I am not sure we have ever looked at the fluorescence of this compound. Would you expect it to be fluorescent?

R. Leif: Yes, there is a reasonable probability since it is a conjugated system, one cannot tell <u>a priori</u>.

N. Kaplan: I think you are correct, but we have never looked at the fluorescence as far as I know. It is a good suggestion and we will investigate the possibility.

S. Weinhouse, Temple University: It occurs to me that the effects which you obtain with these toxins might be an immunogenic effect. It is well known that substances which enhance the immunity of animals will have an effect on transplanted tumors and I wonder if you have looked into this possibility of action?

N. Kaplan: Well, that is why we used the athymic mouse because we wanted to have a system where the immunological factors can be minimized. This is why Gordon Sato and I have gone into the athymic animal so we can look at the pure cytotoxic effects and try to separate them from the immunological effects.

S. Weinhouse: I hade one other point with regard to the streptozotocin. It occurred to me that this substance might act by interfering with acetylglucosamine, either its formation or its incorporation into membrane mucopolysaccharides. Another feature about streptozotocin is that it is perfectly tailored to be a carcinogen and I wonder since it is a nitrosourea derivative if that hazard has been tested or considered.

N. Kaplan: I don't think people working with this compound have looked at its possible carcinogenic activity. Most people have been more interested in the chemotherapy properties of the compound as well as its diabetic-producing effects.

L. Menahan, Medical College of Wisconsin: In your first slide you eluded to NAD requirement for the activation of adenylate cyclase by the cholera toxin and I am particularly interested in this. The toxins usually only activate adenylate cyclase in whole cell preparations rather than the isolated membranes that we conventionally use for hormone studies. What is the protocol? Do you deprive the cells of NAD to show the requirement?

N. Kaplan: There is no question about that. A number of groups now have membrane preparations which are dependent on NAD, and the toxin to induce cyclase activity.

L. Menahan: In other words, the reason people have not found a toxin effect previously, is because they have not added any NAD to their preparation.

N. Kaplan: Well, yes. That is one reason. In most cells there is a great deal of NADase which hydrolyses NAD, in the broken cell preparation, this enzyme becomes active and NAD is destroyed. Gill at Harvard recognized this point and he added nicotinamide to protect the NAD and was able to show from such studies that he could get the cholera toxin effect on the cyclase activity of broken cells.

L. Menahan: What concentration ranges are we talking about, μM or mM?

N. Kaplan: Are you talking about NAD^+? It is in micromolar levels.

L. Menahan: Yes, I was just wondering if it was in the physiological range.

N. Kaplan: Yes, it is in that range for sure.

R. Parks, Brown University: I was intrigued by the prolonged duration of action that you observed with the catecholamines attached, I believe, to a polypeptide?

N. Kaplan: Yes, it is a random copolymer.

R. Parks: The catecholamines are of special interest, because the natural compounds, such as epinephrine and norepinephrine can have their physiological actions terminated by at least three different mechanisms. First of all, they can react with the enzymes monoamineoxidase (MAO), or with catechol-o-methyl transferase (COMT). In addition, in certain tissues, they may be reabsorbed by specific storage granules.

As you are aware, many of the drugs employed for a wide variety of purposes, in fields such as cardiovascular and neuropharmacoloy, interact with one or another of these specific mechanisms. Have you had the opportunity to examine the behavior of your very interesting "bound" catecholamines with any of these inactivating systems?

N. Kaplan: We know that the inactivating system works very poorly on these bound catecholamines. This has been quite well established. The uptake appears not to be a factor with the bound drugs since there is little dissociation from the glass into the free "native" catecholamine.

R. Parks: Have you tried substrate activity with MAO and COMT?

N. Kaplan: Well we have tried these systems and found that the activity is less than 1% of the free compound.

REGULATION OF FATTY ACID BIOSYNTHESIS IN MAMMARY TUMORS

F. AHMAD, P. AHMAD AND D. SCHILDKNECHT
Papanicolaou Cancer Research Institute
Miami, Florida 33123

Abstract:

In an attempt to understand the decreased capacity of mammary tumors to synthesize long chain fatty acids from acyl CoA derivatives, levels of the enzymes involved in this process (acetyl CoA carboxylase and fatty acid synthetase) have been measured in two transplantable mammary tumors (R3230AC and 13762) and compared with those found in the normal lactating mammary gland of the rat. The data obtained show that the assayable levels of both of these enzymes in the mammary tumors are markedly low, i.e. 5-20% of those present in the lactating mammary gland.

Acetyl CoA carboxylase catalyzes the first committed step of this metabolic pathway. Therefore, we have investigated whether the decreased capacity of mammary tumors to synthesize fatty acids is manifest in decreased synthesis of this enzyme and/or is due to the presence of a form(s) of the enzyme that is catalytically less active. For these studies, acetyl CoA carboxylase from the normal gland was resolved to a high degree of purity (7-8 i.u./mg protein) and was used to raise antibodies against it in rabbits. This antiserum was used in the immunotitration of acetyl CoA carboxylase activity of tumorous and of normal gland origins. The data obtained revealed that removal of 50% acetyl CoA carboxylase activity of neoplastic origin required 7-10 fold greater the amount of antisera needed to achieve the same for the enzyme obtained from the normal gland. A comparison of immunological studies contained herein and those reported by Majerus and Kilburn (J. Biol. Chem. 244 (1969) 6254) indicate that there may exist a difference in the factor(s) regulating acetyl CoA carboxylase in mammary tumors as opposed to that in rat hepatomas.

The process of lipogenesis is widely distributed in nature and plays an important physiological role in the normal functioning of various cells. In most of the biological systems studied, the initial steps of this key metabolic pathway are catalyzed by the sequential action of two enzyme complexes, acetyl CoA carboxylase and fatty acid synthetase as shown below.

$$\text{Acetyl CoA} + \text{ATP} + \text{HCO}_3^- \rightleftharpoons \text{malonyl CoA} + \text{ADP} + \text{Pi} \qquad (i)$$

$$\text{Acetyl CoA} + 7 \text{ malonyl CoA} + 14 \text{ NADPH} + 14 \text{ H}^+ \longrightarrow \text{palmitic acid} + 14\text{NADP} + 8 \text{ CoA} + 6\text{H}_2\text{O} + 7 \text{ CO}_2 \qquad (ii)$$

Acetyl CoA carboxylase is a large molecular weight enzyme containing covalently bound biotin which acts as the carboxyl carrier (1). Likewise, fatty acid synthetase is a high molecular weight multienzyme complex in which the covalently bound 4-phosphopantetheine plays a central role in the transfer of acyl groups during elaboration of fatty acids (2,3).

Acetyl CoA carboxylase catalyzes the first step in the sequence of reactions leading to lipogenesis. Therefore, it is not surprising that the activity of acetyl CoA carboxylase of animal origin is regulated by multiple factors: allosteric effectors, inhibition by acyl CoA derivatives of long chain fatty acids, alterations in its rate of synthesis and degradation under differing nutritional conditions and possibly via phosphorylation/dephosphorylation, etc. (1,3-5). Recent investigations on fatty acid synthetase of animal origin reveal that in addition to nutritional manipulations, the hormonal perturbations can regulate the activity of this enzyme by altering the rate of its synthesis or degradation (6).

There is ample evidence suggesting that hepatic cholesterogenesis and lipogenesis respond to dietary fluctuations. However, in the liver neoplasms these processes are unaffected by nutritional manipulations suggesting that the regulation of cholesterogenesis and lipogenesis is deranged in hepatic tumors (7-9).

To elucidate the mechanism of abberant control of fatty acid biosynthesis in liver neoplasms, Majerus and co-workers (4,9) investigated the levels of acetyl CoA carboxylase and fatty acid synthetase in the tumors as well as in the livers of host animals maintained under differing nutritional states. In addition, these workers investigated the properties of partially resolved acetyl CoA carboxylase from the normal rat liver and compared it with the enzyme obtained from a transplantable hepatoma. A comparison of the affinity constants for different substrates, heat inactivation, end-product inhibition, etc., led these workers to suggest that acetyl CoA carboxylase isolated from hepatoma was very similar to the enzyme of normal tissue origin. Therefore, it was proposed that the defect of control of fatty acid biosynthesis in the liver tumors resided in their inability to alter the rate of synthesis or degradation of acetyl CoA carboxylase upon nutritional perturbations rather than due to the presence of structurally modified enzyme (4,9). No conclusive evidence to support this postulate was presented, however.

Mouse mammary adenocarcinomas synthesize long chain fatty acids from acetate and glucose at a markedly reduced rate as compared to the normal lactating mammary gland (10). Whether the mammary tumors contain low levels of the enzymes involved in the elaboration of fatty acids or whether it is due to the presence of modified form(s) of these enzymes needed resolution. To answer this question, we have studied the levels of acetyl CoA carboxylase and fatty acid synthetase in two transplantable mammary tumors of the rat and compared it with those present in the normal lactating mammary gland. Furthermore, we have purified acetyl CoA carboxylase from the normal mammary gland to a high degree of purity and raised antibodies against it in rabbits. The anti-acetyl CoA carboxylase serum has been used to test whether partially resolved preparations of acetyl CoA carboxylase (from both the tumors and the normal lactating mammary gland) contain a constant amount of immunologically precipitable enzyme per unit of activity.

In Table 1, the levels of acetyl CoA carboxylase determined on the high speed supernatant obtained from the normal gland are compared with those found in two transplantable mammary tumors of the rat. The mammary tumors under investigation are 13762 and R3230AC, which have been graded as of fast and slow-medium growth rates, respectively. Both of these tumors were carried in Fischer rats and excised after their size had reached 2-3 cm. Histologically, these tumors have been classified as mammary adenocarcinomas.

TABLE I

LEVELS OF ACETYL CoA CARBOXYLASE IN THE NORMAL LACTATING MAMMARY GLAND AND IN TWO TRANSPLANTABLE MAMMARY TUMORS OF THE RAT*

		Sp. Activity (nmoles HCO_3^- inc./ min/mg prot.)	munits (per g. of tissue
1.	Lactating mammary gland (21 days P.P.)	~81.0	~1,130
2.	R3230AC (18 days)	~1.95	~37.4
3.	13762 (10-13 days)	~1.70	~7.3

*Acetyl CoA carboxylase was assayed by the incorporation of $^{14}CO_2$ into malonyl CoA as detailed by Miller and Levy (11).

The results of Table I show that the assayable levels of acetyl CoA carboxylase (specific activity as well as munits/g. tissue) are markedly lower in the mammary tumors than those found in the lactating mammary gland. Similar data were obtained when the activity of fatty acid synthetase was measured in these tumors and in the lactating mammary gland (Table II).

TABLE II

LEVELS OF FATTY ACID SYNTHETASE IN THE LACTATING MAMMARY GLAND AND IN TWO TRANSPLANTABLE MAMMARY TUMORS OF THE RAT*

	Sp. Activity (nmoles malonyl CoA inc. per min/mg prot.)	nmoles malonyl CoA Inc. per min/g tissue
Lactating mammary gland (21 days P.P.)	10.5-13.0 (∼11.8)	165-220 (∼192.5)
R3230AC (18 days)	0.15-0.5 (∼0.3)	2.85-14.6 (∼8.7)
13762 (10-13 days)	0.5-0.9 (∼0.7)	0.4-2.7 (∼2.0)

* Fatty acid synthetase activity was measured by the incorporation of $2\text{-}^{14}C$-malonyl CoA into the long chain fatty acids in the presence of acetyl CoA and NADPH.

Although the results presented in Tables I and II may explain the decreased capacity of mammary tumors to synthesize long chain fatty acids from acyl CoA derivatives, these data, however, do not resolve the questions raised previously.

TABLE III

EFFECT OF cAMP OR ATP ON THE ACTIVITY OF ACETYL CoA CARBOXYLASE*

	c-AMP Theophylline	ATP	ATP Theophylline NaF
		% Inhibition	
1. Lactating mammary gland			
700g	14	20	33
100,000g	13	4	9
2. R3230AC			
700g	39	44	58
100,000g	56	39	41
3. 13762			
700g	35	45	60
100,000g	60	40	64

* Prior to assaying acetyl CoA carboxylase (11), the extracts prepared from the normal and the neoplastic tissues were incubated for 15 min. at 37° in the presence of the components shown above which were held at concentrations shown below.

ATP	1mM
cAMP	5mM
Theophylline	1mM
NaF	2mM

In view of the recent reports that the activity of hepatic acetyl CoA carboxylase decreases upon phosphorylation (5), we have investigated this phenomenon in the extracts prepared from normal mammary gland as well as from tumors. The results of these studies are summarized in Table III. In these experiments, prior to assaying acetyl CoA carboxylase, both the low as well as the high speed supernatants obtained from homogenization of the tissues under investigation were pre-incubated with either cAMP + theophylline, ATP or ATP + theophylline + NaF. Each of these treatments resulted in marked inhibition (maximal inhibition ~60%) of acetyl CoA carboxylase activity of tumorous origin whereas the activity of the enzyme from the normal gland was inhibited to a lesser degree (4-33%). These results suggest that acetyl CoA carboxylase may be subject to modification (e.g., phosphorylation) with concomitant change in its activity as has been reported in case of the hepatic enzyme (5). This process could be mediated by either cAMP-dependent or cyclic nucleotide-independent protein kinase. Interestingly, not only do the tumors contain lower assayable levels of acetyl CoA carboxylase than the normal lactating mammary gland (Table I), but also the enzyme activity of neoplastic origin appears to undergo change more readily (Table III) by any one of the treatments given in Table III. Using highly purified enzyme, studies are in progress to define more accurately whether the decrease in activity noted above (Table III) results from phosphorylation of acetyl CoA carboxylase.

We have used immunological techniques to test whether in mammary tumors it is the synthesis of acetyl CoA carboxylase that is impaired or whether the enzyme is present in a form that is catalytically less active. For these investigations it was essential to purify the enzyme from the rat lactating mammary gland to a high degree of purity. The purification of acetyl CoA carboxylase was achieved by applying modifications to the method described by Miller and Levy (11). The final step in this purification scheme was chromatography of the partially resolved enzyme over a Biogel A-5m column. This results in an almost complete separation of acetyl CoA carboxylase from the contaminating protein(s). The active protein thus obtained has a specific activity of 7-8 i.u./mg of protein.

We have recently discovered that one of the contaminating proteins which remains associated with acetyl CoA carboxylase through different stages of purification is the fatty acid synthetase of the mammary gland. The fatty acid synthetase separates from acetyl CoA carboxylase during the last step of purification. Perhaps one of the reasons for the separation of fatty acid synthetase from acetyl CoA carboxylase is manifest in the conditions selected to achieve purification of acetyl CoA carboxylase which leads to the dissociation of fatty acid synthetase. For example, prior to gel exclusion chromatography, the protein is exposed to low ionic strength buffer for a substantial duration at 4° to effect solubilization. Both these conditions (low temperature and low ionic strength) are known to dissociate reversibly the fatty acid synthetase to its constituent subunits (12). Fractions containing fatty acid synthetase have been obtained from Biogel A-5m column which upon dialysis against high concentration of phosphate buffer and DTT show a specific activity of > 700 units mg^{-1} of protein (nmoles TPNH oxidized min^{-1}). Polyacrylamide gel electrophoresis of these fractions in the presence of 6M urea + SDS + β-mercaptoethanol reveals > 90% of the protein applied migrates as a single band of $M_r \sim$ 230,000-240,000. Therefore, the scheme developed for the purification of acetyl CoA carboxylase appears to allow concomitant resolution of fatty acid synthetase. Since acetyl CoA carboxylase and fatty acid synthetase appear to co-purify through many steps of purification, it is likely that in the rat mammary gland there exists a complex comprising both of these enzymes. Attempts are being made to isolate such a complex.

Fig. 1.

Polyacrylamide gel electrophoresis of purified acetyl CoA carboxylase. Acetyl CoA carboxylase was denatured in the presence of 6M urea + 1.0% SDS + 1.0% β-mercaptoethanol contained in sodium phosphate buffer, pH 7.0. An aliquot containing 10 μg of protein was subjected to gel electrophoresis on 4% acrylamide gels containing 6M urea + 0.1% SDS + 0.1% β-mercaptoethanol. Coomasie blue was used to stain the protein.

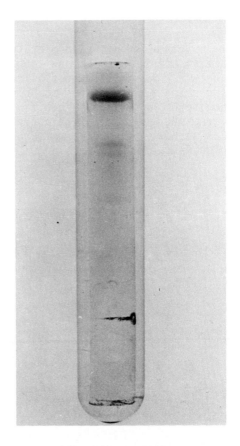

Fig. 1 shows a typical polyacrylamide gel electrophoresis profile of highly purified acetyl CoA carboxylase of rat mammary gland in the presence of 6M urea + 0.1% SDS + 0.1% β-mercaptoethanol. Most of the protein applied to the gel columns migrates as a single band of estimated $M_r \simeq 200,000 - 220,000$. In addition, two faint bands of $M_r \simeq 110,000$ and $125,000$ have been observed in certain preparations (Fig. 1). Whether these bands arise because of an incomplete dissociation of the protomer by these denaturing agents or they are formed as a consequence of limited proteolysis is not yet clear. However, preliminary data obtained upon sedimentation equilibrium of purified enzyme in 6M guanidine HCl + 1mM DTT show a weight homogeneous species of $M_r \simeq 100,000$. Therefore, it appears likely that the rat mammary gland acetyl CoA carboxylase has subunit structure similar to that suggested for the homogeneous rat liver acetyl CoA carboxylase (13).

Antiserum has been prepared against the purified acetyl CoA carboxylase in rabbits. When rat mammary gland acetyl CoA carboxylase of different degrees of purity is subjected to Ouchterlony double diffusion analysis with this antiserum, a single precipitin band is obtained (Fig. 2). These and the data given above support the view that acetyl CoA carboxylase of rat mammary gland has been resolved to a high degree of purity.

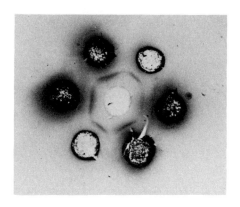

Fig. 2.

<u>Ouchterlony double diffusion analysis of partially resolved acetyl CoA carboxylase from the normal mammary gland.</u> Agar diffusion gels (1.0%) were prepared and developed as described by Nakanishi and Numa (14). Varying amounts of acetyl CoA carboxylase resolved through DE-52 column were applied to the outer wells (20, 40, and 79 munits). The center well contained 95 µg anti acetyl CoA carboxylase serum.

TABLE IV

IMMUNOTITRATION OF ACETYL CoA CARBOXYLASE PARTIALLY PURIFIED FROM THE NORMAL MAMMARY GLAND AND FROM TWO TUMORS OF THE RAT*

Enzyme (Source)	Immune serum required to precipitate 50% of ACC activity (1 munit) (μg)
1. Normal Gland	4
2. R3230AC (Tumor)	40
3. 13762 (Tumor)	28

*Fixed amounts of enzyme of normal and neoplastic tissue purified through the DE-52 stage were added to varying amounts of immune serum. The precipitin reaction mixtures contained in addition final concentrations of 20mM Tris-Cl pH 7.5, 0.15M NaCl, 5mM 2-mercaptoethanol, 1mM EDTA and 20% glycerol. After incubation and 37° for 15 min and storage at 4° overnight, samples were spun at 1500 g for 15 min and the resulting supernatants were assayed for carboxylase activity at 37° by the $^{14}CO_2$ fixation method of Miller and Levy (11).

Table IV shows the results of immunotitration of acetyl CoA carboxylase of normal gland and of tumorous origin which had been carried through identical purification steps (negative adsorption on DE-52) using the antisera prepared against the purified enzyme. It is seen that removal of an equivalent amount (50%) of enzymatic activity of acetyl CoA carboxylase of neoplastic origin requires approximately 7-10 fold greater the amount of immune serum needed to achieve the same for the enzyme obtained from the normal gland. This disproportionate removal of enzymatic activity suggests presence of large amounts of material in the neoplastic tissue which cross-reacts with anti-acetyl CoA carboxylase. The immuno-reactive material in tumors may represent modified form(s) of acetyl CoA carboxylase which is catalytically less active (e.g. phosphorylated enzyme, degraded product of acetyl CoA carboxylase which possesses the ability to bind the antibody or perhaps apo-acetyl CoA carboxylase, etc.).

The results of immunotitration of acetyl CoA carboxylase activity of the rat mammary tumors (Table IV) are in conflict with the data communicated by Majerus and Kilburn on the rat hepatoma enzyme (4). Their preliminary data suggested that the levels of acetyl CoA carboxylase activity in hepatomas paralleled the content of immunologically reactive enzyme in the liver neoplasms (4). Clearly, further investigations are required to resolve this discrepancy.

This investigation is supported by USPH grant #CA-15196.

References

1. M.D. Lane, J. Moss, and S.E. Polakis, Current Topics in Cellular Regulation, eds. B.L. Horecker, and E.R. Stadtman, (Academic Press, Inc. New York, 1974), Vol. 8, p. 139.

2. F. Lynen, Fed. Proc. 20 (1961) 941.

3. P.R. Vagelos, Biochemistry of Lipids, eds. H.L. Kornberg, and D.C. Phillips, (Butterworths, London, England, 1974), Vol. 4, p. 99.

4. P.W. Majerus, and E. Kilburn, J. Biol. Chem. 244 (1969) 6254.

5. C.A. Carlson, and Ki-Han, Kim, Arch. Biochem. Biophys. 164 (1974) 478.

6. M.R. Lakshmanan, C.M. Nepokroeff, and J.W. Porter, Proc. Natl. Acad. Sci. USA, 69 (1972) 3516.

7. M.D. Siperstein, and F.M. Fagan, Cancer Res. 24 (1964) 1108.

8. J.R. Sabine, S. Abraham, and H.P. Morris, Cancer Res. 28 (1968) 46.

9. P.W. Majerus, R. Jacobs, M.B. Smith and H.P. Morris, J. Biol. Chem. 243 (1968) 3588.

10. J.C. Bartley, H. McGrath, and S. Abraham, Cancer Res. 31 (1971) 527.

11. A.L. Miller, and H.R. Levy, J. Biol. Chem. 244 (1969) 2334.

12. S. Smith and S. Abraham, Methods Enzymol. 35 (1975) 65.

13. H. Inoue, and J.M. Lowenstein, J. Biol. Chem. 247 (1972) 4825.

14. S. Nakanishi, and S. Numa, Eur. J. Biochem. 16 (1970) 161.

Discussion

F. Lynen, Max-Planck-Institut für Biochemie: I would like to comment on the experiment about the interconvertibility of acetyl CoA carboxylase. You were very careful in your statement and I agree with you. The observation that incubation with ATP inactivates the enzyme does not prove that inactivation is due to a phosphorylation of the enzyme. It would be essential to find also a reactivation of the inactivated enzyme by incubation in the absence of ATP.

F. Ahmad, Papanicolaou Cancer Research Institute: Very true. We have attempted to partially purify protein kinase from the normal mammary gland. This preparation contains both the cyclic -nucleotide dependent and independent protein kinase when tested with histones. We employed this preparation of protein kinase to test its ability to inactivate partially purified acetyl CoA carboxylase in the presence of ATP. The data indicated marked decrease in the incorporation of radioactivity from $H^{14}CO_3^-$ into malonyl CoA which was proportionate to the amount of protein kinase preparation added. However, closer examination revealed that one of the reasons for decreased incorporation of radioactivity from $H^{14}CO_3^-$ into the end-product (malonyl CoA) was the presence of malonyl CoA decarboxylase in the protein kinase preparation. We are planning to purify protein kinase further in an attempt to separate it from malonyl CoA decarboxylase. The purified protein kinase will be used to test its ability to phosphorylate highly purified acetyl CoA carboxylase.

S. Weinhouse, Temple University: Did you notice, as in many cases where phosphorylation would inactivate an enzyme, that the activity might be restored through dephosphorylation? Have you tested this possibility?

F. Ahmad: No, we have not. Carlson and Kim (Arch. Biochem. Biophys. 164:478, 1974) have reported on the inactivation of hepatic acetyl CoA carboxylase upon phosphorylation. This phosphorylated enzyme could be dephosphorylated upon incubation in the presence of Mg^{++} with concomitant increase in the enzymatic activity. However, these experiments were carried out using somewhat cruder preparations of the enzyme.

B. Cameron, Papanicolaou Cancer Research Institute:
Have you done any sort of inverse immunologic experiments? Have you been able to raise antibodies to enzymes from the tumor cell and study these antibodies?

F. Ahmad: No, we have not purified the enzyme from tumor yet.

M. Rouleau, National Institutes of Health: In all your experiments you compare the ACC and FAS activities in tumors to the ones found in normal lactating mammary gland. Since the activity of these enzymes is increased during lactation, do you have any information on the values of the activity of these enzymes in normal nonlactating mammary gland and why do you use lactating mammary gland as a reference since the 2 tumors are not milk-producing tissues?

F. Ahmad: You mean compare the data with the normal and not the lactating mammary gland. No, we have not done that.

S. Weinhouse: I have another question I would like to ask you. Fatty acid synthetase activity is probably regulated *in vivo* by the availability of NADPH, which has to be generated, I think in the main, through glucose-6-P dehydrogenase. Have you looked into this possibility for control in the mammary tumor?

F. Ahmad: We have not looked at it. However, the results of dietary manipulations suggest that the enzymes involved in the process of liogenesis undergo changes in their levels coordinately. Thus, it is possible that the levels and/or the activities of enzymes responsible for providing NAD(P)H for fatty acid biosynthesis in the mammary tumors are also decreased. At present, we do not have data to support this statement.

SELECTIVE INHIBITION OF THE 3' TO 5' EXONUCLEASE ACTIVITY ASSOCIATED WITH MAMMALIAN DNA POLYMERASE δ [1]

J.J. BYRNES[2], K.M. DOWNEY[3], V. BLACK, L. ESSERMAN and A.G. SO[4]
Department of Medicine, Veterans Administration Hospital and the Departments of Medicine and Biochemistry, University of Miami School of Medicine

Abstract: A new species of DNA polymerase has been highly purified from erythroid hyperplastic bone marrow. This DNA polymerase (δ), in contrast to previously described mammalian DNA polymerases (α, β, γ) is associated with a very active 3' to 5' exonuclease. Similar to the 3' to 5' exonuclease associated with prokaryotic DNA polymerases, this enzyme activity catalyzes the removal of 3'-terminal nucleotides from DNA as well as a template-dependent conversion of deoxynucleoside triphosphates to monophosphates.

Experimental evidence is presented suggesting that the exonuclease activity is an integral part of DNA polymerase δ. The exonuclease activity is not separable from the DNA polymerase activity by ion exchange chromatography, and upon sucrose density gradient centrifugation the two activities co-sediment at either 7S or at 11S depending on the ionic strength. Inhibitors of the DNA polymerase activity such as hemin and Rifamycin AF/013 which act by dissociating the enzyme from the DNA template also inhibit the exonuclease activity.

The 3' to 5' exonuclease activity of this DNA polymerase can be selectively inhibited by 5'AMP or 6-mercatopurine ribonucleoside monophosphate, while the DNA polymerase activity is not affected.

It is proposed that the mutagenicity of 6 mercaptopurine in Escherichia coli may be due to the selective inhibition of the proof reading exonuclease activity associated with DNA polymerase and consequent incorporation of mismatched bases into DNA. Similarly, the inhibition of the 3' to 5' exonuclease associated with mammalian DNA polymerase δ by 6 mercaptopurine ribonucleoside monophosphate may account for its being a carcinogen.

INTRODUCTION

The high fidelity of DNA replication in prokaryotes has been shown to be due to the presence of a 3' to 5' exonuclease associated with DNA polymerase. This exonuclease is an integral part of E. coli DNA polymerase I and is found to be present in all prokaryotic DNA polymerases thus far studied (1). The 3' to 5' exonuclease has been shown to have a proof reading function during DNA synthesis, as it removes mismatched nucleotides during DNA polymerization (2). It was further observed that mutations in the gene coding for T4 DNA polymerase could result in either an increased rate (mutator) or a decreased rate (antimutator) of spontaneous mutation (3). The increased frequency of spontaneous mutation in the mutator strains has been shown to be due to a much higher ratio of DNA polymerase to exonuclease activity (i.e., the exonuclease activity is low or altered and thus unable to remove errors of incorporation) and, conversely, the DNA polymerase purified from the antimutator strain has a much higher nuclease to polymerase ratio (4,5). Thus it is clear that this exonuclease activity plays an important role in replication fidelity and the prevention of mutation.

Fidelity of DNA replication is also rigidly maintained in eukaryotes. However, in contrast to bacterial DNA polymerases, 3' to 5' exonuclease activity is reported not to be associated with mammalian DNA polymerases (6,7,8,9,10). It is now generally believed that all purified mammalian DNA polymerases contain no exonuclease activity (See reviews in 1,8,11,12,13). Since the frequency of spontaneous mutation in mammalian cells is not significantly different from that in bacteria, it has been suggested that a different mechanism must exist for controlling error frequency or mutation in mammalian cells (8,10,13).

We have recently reported the purification of a new species of high molecular weight DNA polymerase from erythroid hyperplastic bone marrow (16). This DNA polymerase (DNA polymerase δ) is similar to the previously reported high molecular weight DNA polymerase α (14,15) in its sedimentation properties but may be separated from DNA polymerase α by chromatography on DEAE-Sephadex and hydroxylapatite. DNA polymerase δ differs from the other mammalian DNA polymerases in its template specificity and its association with a very active 3' to 5' exonuclease activity. We wish to present further data showing that: (1) a nucleolytic activity is associated with DNA polymerase δ; (2) this nucleolytic activity is a 3' to 5' exonuclease; (3) the exonuclease activity can be selectively inhibited by 5'AMP and 6 mercaptopurine ribonucleoside monophosphate without affecting DNA polymerase activity.

EXPERIMENTAL

DNA polymerase δ was prepared from erythroid hyperplastic bone marrow as described by Byrnes et al. (16). Assays for DNA polymerase activity, 3' to 5' exonuclease activity and template dependent generation of monophosphates were as previously described (16). Sucrose density gradient analysis (14) and polyethyleneimine cellulose thin layer chromatography (16) were as reported.

RESULTS AND DISCUSSION

Co-Chromatography of DNA Polymerase with 3' to 5' Exonuclease Activity.

The purification of DNA polymerase δ from the microsomal extract of the erythroid hyperplastic bone marrow is summarized in Table 1. Starting with 200 grams of bone marrow obtained from 50 phenylhydrazine treated rabbits, 600 μg of protein are obtained with a specific activity in excess of 2000 nmoles TMP incorporated per hour per mg of protein. As shown in Fig. 1, two DNA polymerase activities are partially separated on DEAE-Sephadex chromatography when poly d(A-T) and activated calf thymus DNA are used as templates. The polymerase activity that elutes at 0.09 M KCl prefers activated calf thymus DNA over poly d(A-T) as template while the polymerase activity that elutes at 0.13 M KCl has a greater preference for poly d(A-T) as template. The most active fractions with poly d(A-T) as template appear as a trailing shoulder in the activated calf thymus DNA-directed elution profile. The elution profile of the 3' to 5' exonuclease activity, as measured by the release of a 3' terminal [3H]TMP from poly [d(A-T)]-[3H]TMP, corresponds to that of the poly [d(A-T)]-directed DNA polymerase activity, while activated calf thymus DNA directed-DNA polymerase contains little or no associated 3' to 5' exonuclease activity. For this reason we have used poly d(A-T) as template in the purification scheme summarized in Table 1.

Table 1

Purification of DNA Polymerase δ

Step	Total Activity (units)	Protein (mg)	Specific Activity (units/mg of protein)	% Yield
I. Microsomal Extract	4994	2040	2.5	100
II. 60% Ammonium Sulfate Precipitate	4839	1337	3.6	97
III. Phosphocellulose Chromatography	2664	83.7	31.8	53
IV. DEAE Sephadex A-25 Chromatography	1878	15.0	125	38
V. Hydroxylapatite Chromatography	1283	0.6	2138	26

Poly [d(A-T)] was used as template to determine DNA polymerase activity. Assay conditions were as previously described (16).

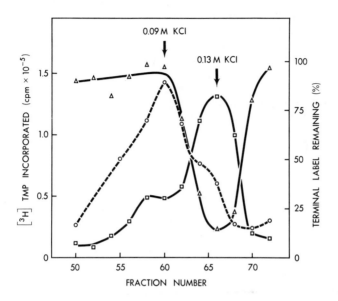

Fig. 1. DEAE-Sephadex A-25 Chromatography of DNA Polymerase.

Experimental details were as described previously (16). O----O DNA polymerase activity assayed with activated calf thymus DNA; □——□ DNA polymerase activity assayed with poly {d(A-T)}; △——△ 3' to 5' exonuclease activity.

The two DNA polymerase activities are further resolved on hydroxylapatite chromatography as shown in Fig. 2. Both the polymerase activity that is active with poly d(A-T) (DNA polymerase δ) and the exonuclease activity elute together at 0.04 M potassium phosphate, whereas the peak of activated calf thymus DNA-directed activity (DNA polymerase α) elutes at 0.075 M potassium phosphate and contains no exonuclease activity. Thus the use of activated calf thymus DNA as template to assay for DNA polymerase activity during purification leads to selective purification of DNA polymerase α containing no exonuclease activity. DNA polymerase δ activity is associated with a small protein peak, while DNA polymerase α is associated with the major protein fraction. The possibility that the activated calf thymus DNA-directed polymerase contains an exonuclease activity but prefers activated calf thymus DNA as substrate was considered. Activated calf thymus DNA labeled at the 3' terminus with [^{3}H]TMP was used to assay for exonuclease activity with both DNA polymerase activities. The pattern of exonuclease activity with activated calf thymus DNA·[^{3}H]TMP as substrate is similar to that with poly{d(A-T)}·[^{3}H]TMP and is associated with DNA polymerase δ and not with DNA polymerase α

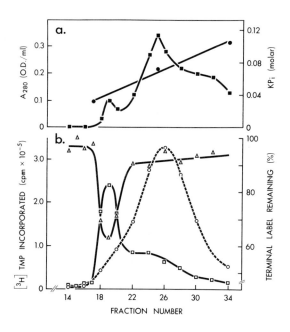

Fig. 2. Resolution of Two DNA Polymerase Activities by Hydroxylapatite Chromatography.

Assay conditions were as described previously (16). a. ■——■ absorbance at 280 nm; ●——● potassium phosphate concentration. b. □——□ DNA polymerase activity assayed with poly{d(A-T)} ; ○·····○ DNA polymerase activity assayed with activated calf thymus DNA; △——△ 3' to 5' exonuclease activity.

Co-sedimentation of DNA Polymerase δ and Its Associated Exonuclease.

To further establish that the 3' to 5' exonuclease activity is an inherent property of DNA polymerase δ, we have further purified the DNA polymerase by sedimentation on sucrose density gradients both at low and high ionic strength. Similar to DNA polymerase α (14,15,8), DNA polymerase δ also exists either as a monomer of 7S at high ionic strength or as a dimer of 11S at low ionic strength. As shown in Fig. 3, in the presence of 0.05 M KCl both DNA polymerase and exonuclease activities sediment together at 11S, while at 0.3 M KCl both DNA polymerase and exonuclease activities cosediment at 7S in a very sharp peak (Fig. 4). This strongly suggests that both DNA polymerase and exonuclease activities are catalyzed by the same protein molecule, as in the case in prokaryotes. The purification of DNA polymerase δ obtained by sedimentation on sucrose density gradients has not been quantitated because of the small amount of very dilute protein and the instability of the enzyme under the conditions of sedimentation.

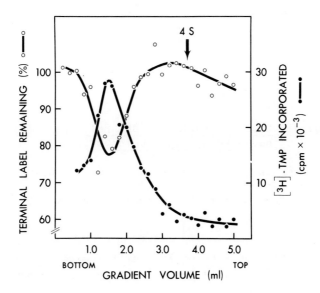

Fig. 3. Cosedimentation of DNA Polymerase δ and Exonuclease at Low Ionic Strength.

Experimental details were as previously described (14). The concentration of KCl was 0.05 M. ●——● DNA polymerase activity; o——o 3' to 5' exonuclease activity.

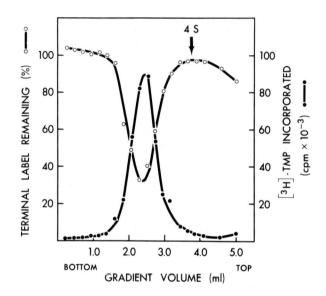

Fig. 4. <u>Cosedimentation of DNA Polymerase δ and Exonuclease at High Ionic Strength.</u>

Experimental details were as previously described (14). The concentration of KCl was 0.3 M. ●——● DNA polymerase activity; o——o 3' to 5' exonuclease activity.

<u>Inhibition Studies Provide Further Evidence that the Exonuclease is an Integral Part of the DNA Polymerase.</u>

We have previously shown that hemin and Rifamycin AF/013 are inhibitors of DNA polymerase activity (14). Moreover, we have demonstrated that the mechanism of inhibition by these two agents involves the dissociation of DNA polymerase from the template (17). We reasoned it would be likely that agents which dissociate the polymerase from polynucleotides would likewise inhibit the exonuclease activity, if the exonuclease is an integral part of the DNA polymerase. To test this postulate the effects of hemin and Rifamycin AF/013 on exonuclease activity and DNA polymerase activity were determined.

As shown in Table 2, hemin at a concentration of 2.4 µM inhibits the DNA polymerase activity 50%, while at the same concentration the 3' to 5' exonuclease activity is inhibited 49%. Similarly Rifamycin AF/013 at a concentration of 20 µg/ml inhibits the DNA polymerase activity 49% and the 3' to 5' exonuclease 35%. These results suggest that both activities reside on the same enzyme molecule, as dissociation of the DNA polymerase from the template/primer would likewise displace the 3' to 5' exonuclease from its substrate, the 3' terminus of the primer.

Table 2

Inhibitors of Both DNA Polymerase and 3' to 5' Exonuclease Activities

Inhibitor	Concentration	Per Cent Inhibition DNA Polymerase Activity	3' to 5' Exonuclease Activity
None	-	0	0
Hemin	2.4 µM	49	50
	6.0 µM	91	82
Rifamycin AF/013	20 µg/ml	47	35
	40 µg/ml	91	86

Assay conditions for DNA polymerase and exonuclease activities were as previously described (16).

Characterization of Nuclease Activity.

That the nuclease activity is directed at the 3' terminus of the template/primer is demonstrated by the rapid release of the labeled 3-terminal nucleotide from poly{d(A-T)}·{^{3}H}TMP as compared to the relatively slow conversion of uniformly labeled {^{3}H}poly d(A-T) to acid solubility (Fig. 5). Incubation of 2 units DNA polymerase with 1.5 nmole poly{d(A-T)}·{^{3}H}TMP or an equal amount of uniformly labeled {^{3}H}poly d(A-T) resulted in 85% release of {^{3}H}TMP from terminally labeled poly{d(A-T)}·{^{3}H}TMP after five minutes of incubation at 37°C, whereas after 30 minutes incubation only 20% of uniformly labeled {^{3}H}-poly d(A-T) was rendered acid soluble. The released product was identified at TMP in both instances by thin layer chromatography of the reaction products on polythylenimine cellulose.

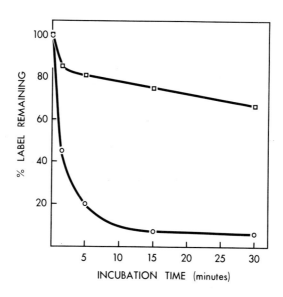

Fig. 5. <u>Hydrolysis of {^{3}H}TMP labeled poly{ d(A-T)}.</u>

Assay conditions were as described previously (16). o——o terminally labeled poly {d(A-T)}·{^{3}H}TMP; □——□ uniformly labeled {^{3}H}TMP - poly {d(A-T)}.

Further evidence for the 3' to 5' exonucleolytic nature of this activity is demonstrated by protection of the 3'-terminus by complementary deoxyribonucleoside triphosphates. The nucleolytic release of the 3' terminal [3H]TMP from poly {d(A-T)}·[3H]TMP is prevented by chain elongation. As illustrated in Fig. 6, in the absence of dATP and TTP, the rate of hydrolysis of [3H]TMP from the 3' terminus of poly {d(A-T)}·[3H]TMP is very rapid, approximately 80% being hydrolyzed in 5 minutes, and complete removal is attained in 15 minutes. Either dATP or TTP alone provides very little protection, since neither nucleotide by itself can extend the DNA chain. However, in the presence of both dATP and TTP, chain elongation can occur and the [3H]TMP is almost completely protected from nucleolytic attack by the 3' to 5' exonuclease. Protection of the 3' terminal nucleotide by DNA synthesis confirms the 3'-exonuclease character of the enzyme and also argues against the presence of endonuclease activity.

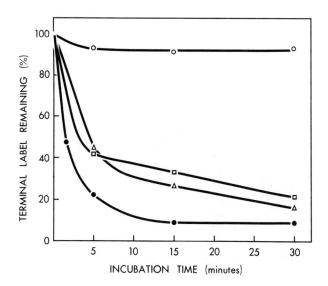

Fig. 6. Effect of Chain Elongation on Hydrolysis of Terminally Labeled poly {d(A-T)}·[3H]TMP.

Assay conditions were as previously reported (16) except: ●——● no dATP or TTP; □——□ 0.16 mM dATP; △——△ 0.16 mM TTP; ○——○ 0.08 mM each dATP and TTP.

Template-dependent Conversion of Nucleoside Triphosphates to Free Nucleoside Monophosphates.

Similar to the 3' to 5' exonuclease of E. coli DNA polymerase I and T4 DNA polymerase (5,18,19), the exonuclease activity associated with marrow DNA polymerase catalyzes the template-dependent conversion of deoxyribonucleoside triphosphates to monophosphates as shown in Table 3. This is a characteristic of 3' to 5' exonuclease activity which reflects alternate incorporation and hydrolysis of the 3' terminal nucleotide of the template/primer (19). In no case was there conversion of nucleoside tri- to monophosphate in the absence of template and the specificity of the reaction for substrates complementary to the template was high. Deoxyribonucleoside triphosphates not complementary to the template (dGTP and dCTP) were converted to monophosphates at a much reduced rate and ATP was likewise converted to AMP to a very limited extent. The fact that a small amount of non-complementary nucleotide is incorporated and released as monophosphate emphasizes the proof reading function of the 3' to 5' exonuclease.

Table 3

Template Requirement and Specificity for Conversion of Nucleoside Triphosphate to Nucleoside Monophosphate.

Template	[3H] Labeled Triphosphate	[3H] Labeled Nucleotide Incorporated into Polymer	Free [3H] Labeled Nucleotide Monophosphate Formed
		pmoles	pmoles
poly [d(A-T)]	dTTP	43	189
none	dTTP	2	2
poly [d(A-T)]	dATP	45	475
none	dATP	2	2
poly [d(A-T)]	dGTP	3.0	13
none	dGTP	1	1
poly [d(A-T)]	dCTP	3.0	3.2
none	dCTP	1.8	1
poly [d(A-T)]	rATP	1	4.7
none	rATP	1	2.0

Assay conditions were as previously reported. Values reported with TTP as substrate were obtained at a concentration of 3.2×10^{-5} M and a specific activity of 19 cpm/pmole; with dATP, 3.2×10^{-5} M and 15 cpm/pmole; with dGTP, 0.4×10^{-5} M and 138 cpm/pmole; with dCTP, 0.12×10^{-5} M and 168 cpm/pmole; with rATP, 0.8×10^{-5} M and 71 cpm/pmole.

The inclusion of 5 units of 5'-nucleotidase in either the exonuclease reaction, where the 3'-terminal nucleotide is released as free monophosphate, or in the template-dependent generation of monophosphate reaction, changed the migration of the labeled product on thin layer chromatography to that of the nucleoside ({3H}-thymidine), indicating that the product of both reactions is a nucleoside 5'-monophosphate.

Selective Inhibition of the 3' to 5' Exonuclease Activity by 5'AMP and its Analogs.

As shown in Fig. 7, increasing concentrations of 5'AMP resulted in an inhibition of the 3' to 5' exonuclease activity as measured by the release of {3H}TMP from the 3'-terminus of poly [d(A-T)]·{3H}TMP, and an apparent stimulation of the DNA polymerase activity, as measured by the incorporation of {3H}TTP into acid insoluble product.

Chromatography of the reaction products of the DNA polymerase assay on PEI-cellulose demonstrated that the apparent stimulation of DNA synthesis was the result of inhibition of the exonuclease activity. As also shown in Fig. 7, the reaction products included both {3H}TMP as polynucleotide, the product of the polymerase reaction, and free {3H}TMP, the product of the exonuclease reaction. Increasing concentrations of 5'AMP resulted in a decrease in the amount of free {3H}TMP formed and a corresponding increase in the amount of {3H}TMP produced as polynucleotide. The total amount of {3H}TMP formed remained constant, suggesting that the DNA polymerase activity was unaffected by 5'-AMP, but rather the selective inhibition of the exonuclease activity by 5'AMP prevented the hydrolysis of newly incorporated {3H}TMP, resulting in an increase in net DNA synthesis.

Various analogs of 5'AMP also inhibit the 3' to 5' exonuclease activity. Fig. 8 shows the effects of increasing concentrations of the active metabolite of 6 mercaptopurine (6MP), 6 mercaptopurine ribonucleoside monophosphate (6-MPR-P) on both the 3' to 5' exonuclease activity and the net synthesis of DNA. Similar to 5'AMP, 6-MPR-P selectively inhibits the exonuclease activity while the DNA polymerase activity is not affected.

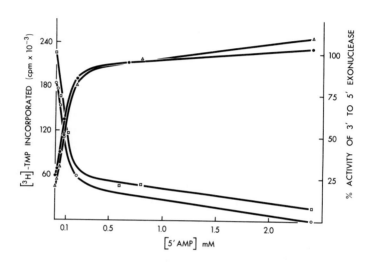

Fig. 7. The Effects of 5'AMP on DNA Synthesis and 3' to 5' Exonuclease Activity.

Assay conditions were as described previously (16) except that in the DNA polymerase assay the concentration of {^{3}H}TTP was 0.16 μM, 50 mCi/μmole. △——△ {^{3}H}TMP incorporated into polynucleotide as measured by standard DNA polymerase assay; ▫——▫ exonuclease activity as measured by the release of {^{3}H}TMP from poly {d(A-T)}·{^{3}H}TMP; ●——● {^{3}H}TMP incorporated into polynucleotide as analyzed by thin layer chromatography on PEI cellulose; ○——○ {^{3}H}TMP as free monophosphate as analyzed by thin layer chromatography on PEI cellulose.

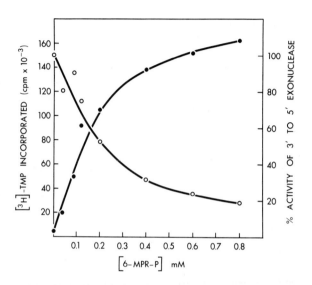

Fig. 8. <u>The Effects of 6-MPR-P on DNA Synthesis and 3' to 5' Exonuclease Activity.</u>

Assay conditions were as previously described (16).
o———o exonuclease activity,   •———• DNA synthesis.

6-mercaptopurine as Mutagen and Carcinogen.

The selective inhibition of the 3' to 5' exonuclease activity by 5'AMP or 6-MPR-P is most interesting, since continued synthesis of DNA with this proof reading mechanism shut off should lead to a higher frequency of misincorporation. Preliminary experiments in our laboratory have shown that the noncomplementary nucleotides dGTP and dCTP are incorporated into poly d(A-T) in the presence of 5'AMP, 6-MPR-P or other purine nucleotides. This is analogous to the mutator strain of T4 bacteriophage in which the ratio of exonuclease activity to polymerase activity is lower than wild type, and the spontaneous mutation frequency is greatly increased (5). Hall and Lehman (4) have shown that the frequency of misincorporation is also increased in in vitro DNA synthesis when the source of the DNA polymerase is a mutator strain of T4.

The ability of 6-MPR-P, the active metabolite of 6-MP, to selectively inhibit the exonuclease activity is especially significant in view of the widespread clinical use of 6-MP and Azathioprine (Imuran), a derivative of 6-MP which is converted to 6-MP in vivo. 6-MP has been shown to be highly mutagenic in E. coli, although the molecular basis for the mutagenicity is not known (20). Studies in our laboratory have shown that 3' to 5' exonuclease activity of E. coli DNA polymerase is also selectively inhibited by 6-MPR-P (manuscript in preparation). This suggests that the mutagenicity of 6-MP in E. coli may be due to the selective inhibition of the proof reading exonuclease activity and consequent incorporation of mismatched bases into DNA.

The carcinogenic effect of 6-MP or Azathioprine has become apparent following its use as an immunosuppresant in renal transplants. It has been estimated that the relative risk of de novo cancer in patients receiving Azathioprine for prolonged periods is 80 times that of the general population (21). The increased incidence of cancer in these patients has been attributed to immunosuppression, however, another possibility is that the carcinogenicity of 6-MP is due to its mutagenicity. The concentration of 6-MP in animal tissues can reach the level of 0.1 mM or greater, a concentration we have found to markedly inhibit the 3' to 5' exonuclease activity of both bacterial and mammalian DNA polymerases. This may be another example of a mutagen being a carcinogen.

SUMMARY

In summary a new species of mammalian DNA polymerase, DNA polymerase δ, has been highly purified from erythroid hyperplastic bone marrow. This DNA polymerase, in contrast to all other mammalian DNA polymerases thus far described, is associated with a very active 3' to 5' exonuclease activity. Our most highly purified DNA polymerase preparation has a specific activity of 2300 nmoles TMP/mg/hr under optimal conditions for DNA synthesis, whereas the exonuclease activity, as measured by template-dependent generation of monophosphate is 380 nmoles/mg/hr with TTP as substrate and 950 nmoles/mg/hr with dATP as substrate.

The 3' to 5' exonucleolytic nature of this enzyme activity is illustrated by the rapid hydrolysis of 3'-terminally labeled poly {d(A-T)}·{^{3}H}TMP as compared to the slow rate of hydrolysis of uniformly labeled {^{3}H}poly d(A-T) and the demonstration that the hydrolytic product is 5'-deoxymononucleotide. That the enzyme is a 3'-exonuclease is further supported by the observation that DNA chain elongation of a 3'-terminally labeled template/primer protects the labeled nucleotide from exonuclease attack, since the labeled nucleotide becomes internally located following DNA chain elongation (2,19).

The data suggests that the 3' to 5' exonuclease activity is an integral part of the DNA polymerase. The DNA polymerase is not separable from the exonuclease by various ion exchange chromatographic procedures, nor by sedimentation in sucrose density gradients both at high and low ionic strength, where the polymerase exists as two molecular weight species.

Further evidence that the exonuclease activity is an inherent property of the DNA polymerase is the inhibition of both activities by hemin and Rifamycin AF/013. These compounds inhibit DNA synthesis by binding to DNA polymerase and causing it to dissociate from the DNA template. At the same concentrations that inhibit DNA polymerase activity 50%, exonuclease activity is also inhibited approximately 50%.

Similar to the E. coli and T4 DNA polymerases (18,19,5), this enzyme also catalyzes a template-dependent conversion of deoxyribonucleoside triphosphates to deoxyribonucleoside monophosphates, a characteristic of 3' to 5' exonuclease activity which reflects alternate incorporation and hydrolysis of the 3'-terminal nucleotide of the template/primer (19)

The 3' to 5' exonuclease activity of DNA polymerase δ is selectively inhibited by 5'AMP and 6 mercaptopurine ribonucleoside monophosphate, while the DNA polymerase activity is not affected. The inhibition of the exonuclease activity prevents the hydrolysis of newly incorporated {^{3}H}TMP and results in an increase in net DNA synthesis.

Since the role of 3' to 5' exonuclease associated with prokaryote DNA polymerase is to insure the fidelity of DNA replication, agents that interfere with this proof reading or editing activity of the DNA polymerase could lead to increased errors of replication and consequent mutation. We

propose that the mutagenicity of 6-mercaptopurine in E. coli
and its carcinogenicity in humans is due to the selective
inhibition of the 3' to 5' exonuclease activity associated
with DNA polymerase resulting in a higher frequency of
misincorporation.

REFERENCES

(1) M. Gefter, Ann. Rev. Biochem. 44 (1975) 45.

(2) D. Brutlag and A. Kornberg, J. Biol. Chem. 247 (1972) 241.

(3) J.F. Speyer, Biochem. Biophys. Res. Commun. 21 (1965) 6.

(4) Z.W. Hall and I.R. Lehman, J. Mol. Biol. 36 (1968) 321.

(5) N. Muzyczka, R.L. Poland and M.J. Bessman, J. Biol. Chem. 247 (1972) 7116.

(6) L.M.S. Chang, J. Biol. Chem. 248 (1973) 6983.

(7) L.M.S. Chang and F.J. Bollum, J. Biol. Chem. 248 (1973) 3398.

(8) F.J. Bollum, Prog. in Nucleic Acid Research and Molecular Biology 15 (1975) 109.

(9) W.D. Sedwick, T.S-F. Wang and D. Korn, J. Biol. Chem. 250 (1975) 7045.

(10) T.S.F. Wang, D. Sedwick and D. Korn, J. Biol. Chem. 249 (1974) 841.

(11) A. Kornberg, DNA Synthesis (W.H. Freeman and Co., 1974).

(12) H.M. Keir and R.K. Craig, Biochemical Society Transactions 1 (1973) 1074.

(13) L.A. Loeb, in: The Enzymes, Vol. 10, ed. P. Boyer (Academic Press, New York, 1974) p. 173.

(14) J.J. Byrnes, K.M. Downey and A.G. So, Biochemistry 12 (1973) 4378.

(15) S. Spadari and A. Weissbach, J. Biol. Chem. 249 (1974) 5809.

(16) J.J. Byrnes, K.M. Downey, V.L. Black and A.G. So, Biochemistry (1976) in press.

(17) J.J. Byrnes, K.M. Downey, L. Esserman and A.G. So, Biochemistry 14 (1975) 796.

(18) M.P. Deutscher and A. Kornberg, J. Biol. Chem. 244 (1969) 3019.

(19) M.S. Hershfield and N.G. Nossal, J. Biol. Chem. 247 (1972) 3393.

(20) S.B. Greer, J. Gen. Microbiol. 18 (1958) 543.

(21) P.S. Schein and S.H. Winokur, Ann. Intern. Med. 82 (1975) 84.

FOOTNOTES

1. Supported by a grant from the National Institutes of Health (NIH AM 09001) and in part by an American Cancer Society Institutional Grant.

2. Veteran Administration Research Associate.

3. Research Career Development Awardee (NIH-KO4-HL-000031).

4. Investigator, Howard Hughes Medical Institute.

Discussion

A. *Mildvan, Institute for Cancer Research*: A critical experiment which I was looking for but did not see, is how much mononucleotide appears when you are carrying out the normal DNA polymerization reaction with saturating amounts of all of the appropriate substrates?

J. *Byrnes, University of Miami*: When the conditions for DNA polymerization are optimal, the monophosphate released is minimal.

A. *Mildvan*: What does minimal mean?

J. *Byrnes*: Minimal means undetectable in our assay system.

A. *Mildvan*: That is the critical question because that tells you that this exonuclease activity is not functioning during the copying of the template. One can calculate for E. coli DNA Polymerase I where we know the fidelity to be as much as 1,000 times too great, if the exonuclease were correcting every mistake, then for every 1,000 correct nucleotides incorporated, we should get 10 free incorrect mononucleotides and this is not observed. This indicates that even in E. coli DNA Polymerase I, an error preventing mechanism is needed in addition to the error correcting mechanism, which operates only under very specialized conditions. We think that the error preventing mechanism might very well be the appropriate orientation of the substrate.

J. *Byrnes*: I believe that the optimal conditions for assaying DNA synthesis in vitro are conditions which shut off the exonuclease activity. In vivo DNA synthesis probably occurs under conditions where both the polymerase and exonuclease are active, and where you would just have net DNA synthesis...some appropriate nucleotide rejected but certainly a large amount of inappropriate nucleotide rejected.

A. *Mildvan*: I completely agree with your comment on the in vivo system. In the in vitro system however, under conditions of optimal DNA synthesis the 3' exonuclease activity cannot be responsible for the high fidelity template copying. Would you agree with the point?

J. Byrnes: Yes. I believe so. However, in experiments we have not described here, we can increase misincorporation of dGTP or dCTP into stable DNA product with poly d(A-T) as template in the presence of a selective inhibitor of the exonuclease. Misincorporation is increased about 5 fold, which is in the range you are talking about from one in a hundred thousand to five in a hundred thousand. It is a very high fidelity none the less.

G. Koch, Roche Institute of Molecular Biology: I enjoyed your fine presentation. I have one question. Did you analyze the peptide composition to look for common peptides in DNA polymerases?

J. Byrnes: No, we have not.

G. Koch: It would be interesting to see if your enzyme shares polypeptide with one of the other DNA polymerases.

J. Byrnes: That is true, we would like to pursue this question, however we are limited by the small amount of enzyme and peptide analysis is not feasible.

G. Koch: However, I think critical for such an analysis would be the use of SDS gel.

J. Byrnes: Do you mean to see if there is a common subunit.

G. Koch: Yes, to see if you have a peptide which is shared by one of the other polymerases.

J. Byrnes: At the present stage of purification there are two bands in native disc gel electrophoresis, and for this reason we have not yet determined SDS gel patterns.

MYELOPEROXIDASE-MEDIATED CYTOTOXICITY[1,2]

S.J. KLEBANOFF, R.A. CLARK[3] and H. ROSEN[4]
Departments of Medicine and Microbiology
University of Washington School of Medicine
Seattle, Washington 98195

Abstract: Phagocytosis by neutrophilic polymorphonuclear leukocytes (PMNs) is associated with a sequence of morphological and biochemical events geared to the concentration within the phagocytic vacuole of agents toxic to the ingested organisms. Associated with these events is the leakage or secretion of toxic agents into the extracellular fluid with the potential for adjacent tissue damage. Among the potentially toxic agents are H_2O_2 acting nonenzymatically or in conjunction with myeloperoxidase (MPO) and a halide, the superoxide anion, hydroxyl radicals, singlet oxygen, lysozyme, proteases or granular cationic proteins. This paper will deal largely with the cytotoxic effect of the MPO-H_2O_2-halide system. The isolated MPO-H_2O_2-halide system is toxic to a variety of mammalian cell types including spermatozoa, erythrocytes, leukocytes, platelets and tumor cells. Phagocytosis by PMNs is also associated with the destruction of adjacent tumor cells. This effect is dependent on a halide and is

---

[1] Abbreviations: CGD - chronic granulomatous disease; DABCO - 1,4 diazobicyclo (2,2,2) octane; MPO - myeloperoxidase; PMN - polymorphonuclear leukocyte; SOD-superoxide dismutase.

[2] Supported by U.S. Public Health Service Grants AI07763, HD02266 and CA18354.

[3] Recipient of Research Career Development Award CA00164 from the National Cancer Institute, DHEW.

[4] Recipient of Postdoctoral Research Fellowship CA05225 from the National Cancer Institute, DHEW.

inhibited by azide, cyanide and catalase. It is not observed when leukocytes which lack MPO or H_2O_2 are employed. These results suggest the release of MPO and H_2O_2 during phagocytosis by PMNs and their interaction with a halide in the extracellular fluid to form a cytotoxic system. The cytotoxicity of the MPO-H_2O_2-iodide system is associated with the iodination of cellular proteins. When chloride is the halide employed, the MPO-mediated system emits light. The formation of singlet oxygen by this system is suggested by the conversion of diphenylfuran to cis-dibenzoylethylene, a reaction believed to be specific for singlet oxygen. The inhibition of cytotoxicity by the singlet oxygen scavengers, 1,4 diazobicyclo (2,2,2) octane (DABCO) and diphenylfuran, is compatible with the involvement of singlet oxygen.

Polymorphonuclear leukocytes (PMNs) respond to microbial invasion by a complex series of events which culminate in the ingestion and destruction of the invading organisms (1). These include the following. 1. The migration of the leukocytes to the locus of invasion under the stimulus of chemotactic factors. 2. The adherence of the organism to the leukocytic plasma membrane and the internalization of that membrane with the incorporation of the organism into an intracellular vacuole. 3. The fusion of adjacent intracellular granules to the phagosome and the rupture of the connecting membrane with the discharge of granular contents into the vacuolar space. The granular contents include myeloperoxidase (MPO), cationic proteins, lysozyme, lactoferrin and a number of acid hydrolases. 4. A burst of leukocytic oxidative metabolism. 5. The destruction of the ingested organisms.

Associated with these events designed to concentrate within the phagocytic vacuole agents toxic to the ingested organism, is the release of these agents to the outside with the potential for extracellular toxicity. Several mechanisms for the release of toxic PMN constituents have been demonstrated (2). Cell lysis due to the action of toxins or other extracellular factors can occur with nonselective release of all cellular components. Rupture of secondary lysosomes, as for example following the ingestion of urate crystals, can release toxic agents into the cytoplasm proper with the lysis of the cell from within. Phagocytosis is associated with the leakage of granular constituents as well as of certain products of the respiratory burst to the outside. Exposure of PMNs to appropriately opsonized but non-ingestable structures such as a large membrane surface coated with immunoglobulin and complement, may result in a similar release of granular constituents (3). Certain humoral mediators such as complement-derived chemotactic factors (4) also can trigger this secretory process. Among the toxic components of the granulocyte which may contribute to the extracellular damage are cathepsins (5), neutral proteases (6, 7), cationic proteins (8, 9), oxygen radicals, H_2O_2 and the MPO-mediated antimicrobial system. This paper will deal with the extracellular cytotoxic effect of PMNs with particular regard to the role of the MPO-mediated system.

THE MYELOPEROXIDASE-MEDIATED ANTIMICROBIAL SYSTEM

MPO, H_2O_2 and a halide form a potent antimicrobial system effective against a variety of microorganisms (for review and extensive bibliography see 10). MPO is present in the neutrophil in very high concentrations. It is present exclusively in the azurophil (or primary) granules of the resting cell, and its discharge into the phagocytic vacuole is readily apparent in electron micrographs of thin sections stained for peroxidase.

The H_2O_2 requirement may be met by the H_2O_2 generated by the leukocyte during the phagocytosis-induced respiratory burst or by the H_2O_2 formed by certain ingested organisms. Phagocytosis is associated with a marked increase in oxygen consumption, and much of this extra oxygen consumed is converted to H_2O_2. The superoxide anion, which is also generated by PMNs during phagocytosis, may serve as an intermediate in H_2O_2 formation by the intact cell. Certain bacteria, e.g., pneumococci, streptococci, lactobacilli, secrete significant amounts of H_2O_2 into the medium. The H_2O_2 so formed is toxic to the organism which formed it or in mixed cultures to an adjacent non-H_2O_2 generating organism, particularly when MPO and a halide are also present. The H_2O_2 formed by microorganisms may contribute significantly to the microbicidal activity of the PMN, particularly when leukocytic H_2O_2 generation is defective as in chronic granulomatous disease (CGD).

Of the cofactors, chloride is present in the leukocyte at a concentration which is considerably greater than that required as a component of the MPO-mediated antimicrobial system. Iodide is approximately 100 times as effective as chloride on a molar basis. Its concentration in serum however is very low. Iodide anions can be replaced by the iodinated hormones, thyroxine and triiodothyronine, as the cofactor in the isolated MPO-mediated antimicrobial system due in part to their deiodination. The thyroid hormones are deiodinated by intact phagocytosing leukocytes and thus may be an additional source of iodide for the intact cell.

The pH optimum of the MPO-mediated antimicrobial system is distinctly acid as is the pH of the vacuolar fluid. The MPO-mediated antimicrobial system is inhibited by catalase, excess H_2O_2 as well as by a number of low molecular weight reducing agents of possible physiological significance, e.g., ascorbic acid, glutathione. Activity of the system therefore may be influenced by the introduction of these inhibitors into the phagocytic vacuole from the leukocyte or with the ingested organism.

CYTOTOXICITY OF THE ISOLATED MPO SYSTEM

The isolated MPO-H_2O_2-halide system is toxic to a variety of cell types other than microorganisms. These include spermatozoa, erythrocytes, leukocytes, platelets and tumor cells.

Spermatozoa. The first demonstrated cytotoxic effect of the MPO-H_2O_2-halide system on mammalian cells was against bull spermatozoa. Myeloperoxidase, lactoperoxidase or uterine peroxidase, when combined with either iodide or thiocyanate ions, had sperm-inhibitory activity as measured by a loss of motility and by a decrease in pyruvate oxidation (11). The H_2O_2 required could be generated by Lactobacillus acidophilus, an organism commonly found in the vaginal canal or by spermatozoal metabolism (12). H_2O_2 formation by spermatozoa was due to the presence of an L-amino acid oxidase with specificity for aromatic amino acids.

Erythrocytes. Erythrocytes are hemolyzed by MPO, a H_2O_2-generating system (glucose + glucose oxidase; xanthine + xanthine oxidase) and either chloride, iodide, thyroxine or triiodothyronine (13). The combined effect of chloride and either iodide or the thyroid hormones is greater than additive. MPO can be replaced by lactoperoxidase in all but the chloride-dependent system. When the xanthine oxidase system is employed as a source of H_2O_2, hemolysis is increased by superoxide dismutase (Table 1).

TABLE 1

Effect of superoxide dismutase on hemolysis by
the myeloperoxidase system*

Supplements	Hemolysis (%)
Xanthine + xanthine oxidase + MPO + Cl^- + I^-	27.4
SOD added	62.0†
SOD + catalase added	0 †
Glucose + glucose oxidase + MPO + Cl^- + I^-	63.1
SOD added	62.4

* Taken from (13).
† Significantly different from complete MPO system, $p < 0.001$, t-test.

Leukocytes. MPO in combination with H_2O_2 and a halide can damage its cell of origin, the PMN. Human PMNs were prepared in high purity (95-98%) by the ficoll-hypaque gradient technique (14), labeled with sodium [^{51}Cr] chromate, washed and suspended in 0.1 M sodium phosphate buffer, pH 7.0 at 10^7 cells/ml. After a one hour incubation at 37°C with the components of the MPO system, release of ^{51}Cr was determined and expressed as a percent of maximum releasable activity (15). As shown in Table 2, cytotoxic activity was dependent on enzymatically active MPO, H_2O_2 and either chloride or iodide. H_2O_2 could be replaced by glucose + glucose oxidase. Edelson and Cohn (16) have demonstrated a cytotoxic effect on human blood mononuclear cells (mixed monocytes and lymphocytes) of a system comprised of lactoperoxidase, glucose, glucose oxidase and iodide.

Platelets. Exposure of platelets to the MPO system results in the release of the biological mediator 5-hydroxytryptamine (serotonin) from the platelet storage granules. Human platelet rich plasma was incubated with ^3H-serotonin, the labeled platelets were washed, suspended in 0.1 M sodium phosphate buffer, pH 6.5 at 10^8/ml and incubated with the components of the MPO system at 37°C for 10 minutes.

TABLE 2

Cytotoxicity of the peroxidase system on human PMNs*

Supplements	^{51}Cr percent release†	
	Chloride	Iodide
MPO + H_2O_2 + halide	61.8^Π	69.5^Π
MPO omitted	-10.4	-3.6
H_2O_2 omitted	-13.7	-2.7
Halide omitted	1.6	1.6
MPO heated	-8.2	2.7
MPO + glucose + glucose oxidase + halide	83.0^Π	69.7^Π

* The reaction mixture contained 0.03 M sodium phosphate buffer pH 7.0, 1.5×10^{-3} M KH_2PO_4, 1.5×10^{-3} M $MgSO_4$, 5×10^5 PMNs and the supplements as follows: MPO, 16 mU (13); H_2O_2, 5×10^{-5} M; NaCl, 0.1 M; NaI, 4×10^{-4} M; glucose, 10^{-2} M; glucose oxidase, 140 mU. Na_2SO_4 0.067 M, was added to maintain isosmolarity to all reaction mixtures which did not contain NaCl. MPO was heated at 100°C for 15 minutes where indicated. Total volume was 0.5 ml.

† Mean of 3-9 experiments calculated as follows:
^{51}Cr percent release = $\frac{\text{Sample cpm-background cpm}}{\text{Maximum cpm-background cpm}} \times 100$
where background tubes contained none of the supplements and maximum tubes contained 1% Triton X-100 (15).

Π Significantly increased above controls, $p < 0.001$, t-test.

Following centrifugation, the amount of ^{3}H released into the supernatant fluid was determined by liquid scintillation counting and data were expressed as a percent of maximum releasable activity. Table 3 demonstrates the release of ^{3}H-serotonin which is dependent on enzymatically active MPO, H_2O_2 and chloride and is blocked by azide, cyanide and catalase, but not by heated catalase. This release may represent a cytotoxic effect on the platelets or could be due to a triggering of the platelet release reaction (17). The concomitant release of small amounts of ^{14}C-adenine in double-label experiments suggests that at least a part of the serotonin release is related to a cytotoxic effect. Nonenzymatic effects of H_2O_2 on platelet aggregation have also been recently demonstrated (18).

TABLE 3

Peroxidase-mediated serotonin release from human platelets*

Supplements	^{3}H-serotonin release†
MPO + H_2O_2 + chloride	57.0
MPO omitted	4.7
H_2O_2 omitted	0.7
Chloride omitted	0.2
MPO heated	2.7
Azide added	3.1
Cyanide added	4.5
Catalase added	2.0
Heated catalase added	60.8

* The reaction mixture contained 0.03 M sodium phosphate buffer pH 6.5, 1.5×10^{-3} M KH_2PO_4, 1.5×10^{-3} M $MgSO_4$, 5×10^6 platelets and the supplements as follows: MPO, 4 mU; H_2O_2, 2×10^{-5} M; NaCl, 0.1 M; sodium azide, 10^{-4} M; sodium cyanide, 10^{-4} M; catalase, 200 U. Na_2SO_4 0.067 M, was added to all reaction mixtures which did not contain NaCl. MPO or catalase was heated at 100°C for 15 minutes where indicated. Total volume was 0.5 ml.

† Mean of 6-12 experiments calculated as in Table 2.

TABLE 4

Cytotoxicity of the xanthine-xanthine oxidase-myeloperoxidase-chloride system on tumor cells*

Supplements	^{51}Cr percent release[†]
Xanthine + xanthine oxidase + MPO + chloride	77.7 ± 3.1 (13)
Xanthine omitted	2.6 ± 4.5 (5)
Xanthine oxidase omitted	-6.3 ± 1.4 (5)
MPO omitted	4.4 ± 3.7 (5)
Chloride omitted	1.3 ± 3.1 (5)
Xanthine oxidase heated	-1.6 ± 0.8 (3)
MPO heated	2.2 ± 0.7 (3)
Catalase added	1.0 ± 1.7 (4)
Heated catalase added	65.8 ± 5.9 (3)
Superoxide dismutase added	81.9 ± 2.9 (8)

* The reaction mixture contained 0.02 M sodium phosphate buffer pH 7.0, 1.5×10^{-3} M KH_2PO_4, 1.5×10^{-3} M $MgSO_4$, 10^{-4} M EDTA, 10^5 tumor cells and the supplements as follows: xanthine, 10^{-4} M; xanthine oxidase, 20 mU; MPO, 16 mU; NaCl, 0.1 M; catalase, 2100 U; superoxide dismutase, 29 U (2.5 µg). Na_2SO_4 0.067 M, was added to all reaction mixtures which did not contain NaCl. MPO, xanthine oxidase and catalase were heated at 100°C for 15 minutes where indicated. Total volume was 0.5 ml.

[†] Mean ± SE of (n) experiments calculated as in Table 2.

Tumor cells. Recent studies by ourselves (15) and others (16, 19) have demonstrated the ability of peroxidases to damage mammalian tumor cells. Edelson and Cohn (16) demonstrated the cytotoxic effect of the peroxidase (lactoperoxidase, myeloperoxidase)-glucose-glucose oxidase-halide system on a line of mouse lymphoma cells designated L1210 using the ^{51}Cr release assay. Philpott et al. (19), utilized specific anti-tumor antibody covalently linked to glucose oxidase to localize H_2O_2 production to the vicinity of the target cell. The H_2O_2 so formed was utilized to kill the tumor cells in the presence of lactoperoxidase and iodide. Schultz et al. (20), have reported the repression of tumor growth in vivo by a combination of MPO and thiotepa. In our studies (15), a line of mouse ascites lymphoma cells designated LSTRA was exposed to the MPO-H_2O_2-halide system, and the cytotoxic effect detected by ^{51}Cr release, trypan blue exclusion, inhibition of glucose C-1 oxidation and loss of oncogenicity for mice. MPO could be replaced by lactoperoxidase, H_2O_2 by glucose and glucose oxidase or by H_2O_2 generating bacteria such as pneumococci or streptococci, and the halide requirement could be met by chloride, iodide or the iodinated hormones, thyroxine and triiodothyronine (15). Table 4 demonstrates the cytotoxic effect of the xanthine-xanthine oxidase-MPO-chloride system on LSTRA cells and the requirement for each component of the cytotoxic system. A requirement for H_2O_2 was indicated by the inhibitory effect of catalase. In contrast superoxide dismutase was not inhibitory. Thus under

TABLE 5

Comparison of the cytotoxic activity of the myeloperoxidase system and granular cationic proteins*

Cytotoxic system	^{51}Cr percent release[†]
MPO (5 µg/ml) + H_2O_2 + chloride	38.5 ± 6.5 (7)
Cationic proteins (50 µg/ml)	45.9 ± 8.4 (11)
Cationic proteins (10 µg/ml)	0.9 ± 0.9 (5)

* The reaction mixture contained 0.07 M sodium phosphate buffer pH 7.0, 1.5x10^{-3} M KH_2PO_4, 1.5x10^{-3} M $MgSO_4$; 10^5 tumor cells, water to a final volume of 0.5 ml and either the MPO system (MPO 5 µg/ml [24 mU/ml]; H_2O_2, 5x10^{-5} M; NaCl, 0.1 M) or cationic proteins at the concentrations indicated. The cationic proteins were a mixture of proteins 3 and 4 (22).

[†] Mean ± SE of (n) experiments calculated as in Table 2.

the conditions employed, the direct toxicity of the superoxide anion appeared to be weak when compared to the H_2O_2 formed in part from it, when combined with MPO and chloride.

Strongly cationic proteins located in the PMN granules have been implicated in the microbicidal activity of the cell (21). These granular cationic proteins have been isolated in highly purified form from human leukemic leukocytes and shown to exhibit microbicidal activity in vitro (22, 23). They also have cytotoxic activity. Table 5 compares the cytotoxic effect against mammalian tumor cells of the MPO-H_2O_2-chloride system to that of purified granular cationic proteins. MPO at a concentration of 5 µg/ml had a cytotoxic effect comparable to that of the cationic proteins at a concentration of 50 µg/ml.

CYTOTOXIC EFFECT OF INTACT PMNS

The ingestion of pre-opsonized zymosan by intact PMNs

TABLE 6

Cytotoxicity of intact PMNs mediated by the peroxidase system

Supplements	^{51}Cr percent release*		
	Normal	CGD	MPO def.
PMN + Zymosan + Cl^- + I^-	26.4	-5.9	-11.1
Azide added	-3.4		
Cyanide added	-8.6		
Catalase added	-6.8		
Heated catalase added	16.2		
H_2O_2 added	27.8	11.6	-2.6
Glucose + glucose oxidase added	26.6	20.9	-5.3
MPO added	26.9	-2.5	36.1

* Taken from (24) with results recalculated as described in Table 2.

results in the destruction of adjacent tumor cells (24, Table 6). This effect is dependent on a halide with the combined effect of chloride and iodide being greater than additive. It is completely inhibited by azide, cyanide and catalase suggesting a role for MPO and H_2O_2. This is further supported by the finding that PMNs which lack MPO (hereditary MPO deficiency) or H_2O_2 formation (chronic granulomatous disease) are not cytotoxic under these conditions. Activity is restored to the MPO-deficient leukocytes by the addition of MPO and to CGD leukocytes by the addition of H_2O_2 or the glucose oxidase system. These studies suggest that phagocytosing PMNs release MPO and H_2O_2 into the extracellular fluid where they interact with a halide to form a cytotoxic system (Figure 1). Such a sequence of events may be operative in vivo, for example, in areas of microbial invasion.

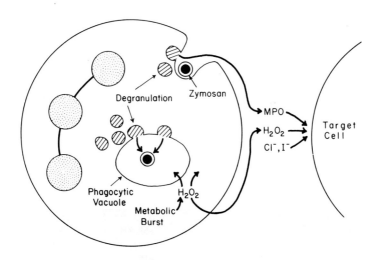

Figure 1. Mechanism of cytotoxicity by PMNs during phagocytosis.

MECHANISM OF THE CYTOTOXICITY OF THE PEROXIDASE SYSTEM

The mechanism of action of the MPO-mediated cytotoxic system is complex (see 10) and will not be considered in detail here. It is probable that MPO, H_2O_2 and the halide interact to produce a toxic agent or agents which affect the cell at multiple sites and in a number of ways which vary with the halide and the characteristics of the cell surface. Only two of these mechanisms will be considered here, halogenation and singlet oxygen formation.

Halogenation. Iodination by lactoperoxidase H_2O_2 and iodide has been employed to identify exposed proteins on the erythrocyte cell membrane (25, 26). When the iodinating con-

TABLE 7

Iodination of tumor cells*

Supplements	Iodination (%)
MPO + H_2O_2 + chloride + tumor cells	17.9 ± 1.5 (4)[†]
MPO omitted	0.6 ± 0.04 (4)
H_2O_2 omitted	0.6 ± 0.06 (4)
Chloride omitted	2.0 ± 0.2 (4)
Tumor cells omitted	2.1 ± 1.3 (3)
MPO heated	0.6 ± 0.1 (4)
Tumor cells heated	35.2 ± 0.7 (2)
Azide added	0.7 ± 0.04 (4)
Cyanide added	1.2 ± 0.3 (4)

* The reaction mixture contained 0.03 M sodium phosphate buffer pH 7.0, 1.5×10^{-3} M KH_2PO_4, 1.5×10^{-3} M $MgSO_4$, 2×10^{-6} M sodium iodide (0.05 µCi ^{125}I) and the supplements as follows: MPO, 40 mU; H_2O_2, 10^{-4} M; NaCl, 0.1 M; tumor cells, 10^7; sodium azide, 10^{-4} M; sodium cyanide, 10^{-4} M. Na_2SO_4 0.067 M, was added to all reaction mixtures which did not contain NaCl. MPO and tumor cells were heated at 100°C for 15 minutes where indicated. Total volume was 0.5 ml. Iodination was determined as previously described (13).

[†] Mean ± SE of (n) experiments.

ditions were mild, little or no hemolysis was observed. However, when the initial iodide concentration was high and the total number of iodine atoms bound was correspondingly high, more extensive membrane damage with lysis occurred (13, 16, 27). Hemolysis by the MPO-H_2O_2-iodide system also was associated with iodination of erythrocyte components (13). With this enzyme both iodination and hemolysis were increased by chloride ions. Exposure of the mouse ascites lymphoma line to the MPO-mediated cytotoxic system also resulted in the iodination of cellular components (Table 7). Each component of the system (MPO, H_2O_2, chloride, tumor cells) were required for optimum iodination under the conditions employed and activity was abolished by the addition of azide or cyanide or by heat inactivation of MPO. When the tumor cells were preheated, however, iodination was increased. This may reflect the greater availability of iodine acceptor sites in denatured proteins or the heat inactivation of inhibitors of iodination (e.g., catalase).

It should be emphasized that the association of iodination with the cytotoxicity of the iodide-dependent system does not necessarily imply a cause-effect relationship. Iodination may be an innocent bystander to the cytotoxic process or, more probably, be one of a number of factors which contribute to cell death.

Singlet oxygen formation. Chloride is oxidized by MPO and H_2O_2 to form a product with strong oxidative capacity which resembles hypochlorite in its chemical properties (28). Hypochlorite reacts with H_2O_2 to produce a weak red chemiluminescence due to the formation of singlet molecular oxygen (29). The hypochlorite formed by the MPO system therefore might be expected to react with excess H_2O_2 to form singlet oxygen, and the chemiluminescence observed during the interaction of MPO, H_2O_2 and a halide (30, 31) is compatible with singlet oxygen formation. Table 8 demonstrates the light emission generated by MPO, H_2O_2 and chloride and the requirement for each component of the system. When zymosan, a yeast cell wall preparation commonly employed as a phagocytic particle, is added light emission is increased, and under these conditions chemiluminescence is only partially decreased by the deletion of chloride. This emphasizes that light emission is not necessarily the result of singlet oxygen decay but may be due to formation of another excited species. This species may be dependent on singlet oxygen for its formation or may be formed entirely independently of singlet oxygen, for example, by the oxidation of a zymosan component by MPO and H_2O_2.

TABLE 8

Chemiluminescence by the myeloperoxidase system*

Supplements	Chemiluminescence (cpm)
MPO + H_2O_2 + chloride	3,900
MPO omitted	200
H_2O_2 omitted	500
Chloride omitted	600
MPO + H_2O_2 + chloride + zymosan	13,900
MPO omitted	200
H_2O_2 omitted	500
Chloride omitted	9,800

* The reaction mixture consisted of 0.01 M sodium acetate buffer pH 5.0 and where indicated 80 mU MPO, 10^{-4} M H_2O_2, 0.1 M NaCl and 1 mg zymosan. Total volume was 1.0 ml. Chemiluminescence was measured in a liquid scintillation counter (out of coincidence mode; single photomultiplier tube activated; 0 to ∞ window) during the initial 2 minutes of incubation.

More direct evidence for the formation of singlet oxygen by MPO, H_2O_2 and chloride is the conversion by this system of diphenylfuran to cis-dibenzoylethylene (Table 9), a reaction which is believed to be specific for singlet oxygen (32). The formation of cis-dibenzoylethylene by the MPO-glucose-glucose oxidase-chloride system was indicated by separation of the products by thin layer chromatography and comparison of the spots to authentic standards, either under ultraviolet illumination or following spraying with 0.5% dinitrophenyl-hydrazine in 2N HCl. Quantitation of the conversion was determined by elution of the spot with ethanol and comparison of the absorbancy at 260 nm to a standard curve prepared from synthesized cis-dibenzoylethylene. Each component of the MPO-glucose-glucose oxidase-chloride system was required and cis-dibenzoylethylene formation was inhibited by azide and cyanide, by the singlet oxygen quencher 1,4-diazobicyclo (2,2,2) octane (DABCO) and by catalase.

TABLE 9

Conversion of diphenylfuran to cis-dibenzoylethylene by the myeloperoxidase system*

Supplements	cis-Dibenzoylethylene (nmoles)
MPO + glucose + glucose oxidase + chloride	31.5
MPO omitted	3.0
Glucose omitted	2.5
Glucose oxidase omitted	2.0
Chloride omitted	3.5
Azide added	1.0
Cyanide added	1.5
DABCO added	6.0
Catalase added	1.5
Heated catalase added	26.0

* The reaction mixture contained 0.05 M sodium acetate buffer pH 5.0, 2.5×10^{-5} M diphenylfuran (50 nmoles) and the supplements as follows: MPO, 16 mU; glucose, 10^{-2} M; glucose oxidase, 25 mU; NaCl, 0.1 M; sodium azide, 10^{-3} M; sodium cyanide, 10^{-3} M; catalase, 850 U. Final volume was 2.0 ml. Catalase was heated at 100°C for 15 minutes where indicated.

Singlet oxygen can dissipate its excess energy not only by the emission of light as it reverts to the triplet ground state but also by participation in oxygenation reactions. That reactions of this type on the surface of the cell may be toxic is suggested by the photodynamic action of dyes. This toxicity due to dye-sensitized photooxygenation reactions is

TABLE 10

Inhibition of the cytotoxic activity of the myeloperoxidase system by singlet oxygen quenchers*

Supplements	^{51}Cr percent release†
DABCO	3.8 ± 1.3 (8)
Diphenylfuran	9.6 ± 2.9 (5)
MPO + glucose + glucose oxidase + chloride	59.6 ± 2.3 (6)
DABCO added	20.6 ± 4.8 (5)^{II}
Diphenylfuran added	30.7 ± 4.4 (5)^{II}

* The reaction mixture contained 0.04 M sodium phosphate buffer pH 7.0, 1.5×10^{-3} M KH_2PO_4, 1.5×10^{-3} M $MgSO_4$, 10^5 tumor cells and the supplements as follows: MPO, 4 mU; glucose, 2×10^{-3} M; glucose oxidase, 62 mU; NaCl, 0.1 M; DABCO, 10^{-3} M; diphenylfuran, 10^{-5} M. Total volume was 0.5 ml.

† Mean ± SE of (n) experiments calculated as in Table 2.

^{II} Significantly different from complete MPO system, $p < 0.001$, t-test.

believed to be mediated by singlet oxygen. That singlet oxygen may also be responsible in part for the toxicity of the peroxidase system is suggested by the inhibitory effect of singlet oxygen quenchers. The inhibitory effect of DABCO and diphenylfuran on the cytotoxic activity of the MPO-glucose-glucose oxidase-chloride system is shown in Table 10. Both DABCO and diphenylfuran at concentrations greater than those employed in Table 10 had a cytotoxic effect on the tumor cells in the absence of the peroxidase system.

In summary, these studies demonstrate a potent cytotoxic effect of myeloperoxidase, H_2O_2 and a halide on a variety of cell types including mammalian tumor cells, and evidence is presented which suggests that this system is involved in the cytotoxic effect of intact phagocytosing PMNs. Among the mechanisms considered was halogenation of exposed proteins on the target cell surface and the formation of singlet molecular oxygen. The formation of this excited species by the MPO-H_2O_2-chloride system was indicated by the conversion of diphenylfuran to cis-dibenzoylethylene and the involvement of singlet oxygen in the toxicity was suggested by the inhibitory effect of the singlet oxygen scavengers, DABCO and diphenylfuran.

REFERENCES

(1) J.G. Hirsch, in: The Inflammatory Process, Vol. 1, 2nd edition, eds. B.W. Zweifach, L. Grant and R.T. McCluskey (Academic Press, New York, 1974) p. 411.

(2) G. Weissmann, R.B. Zurier and S. Hoffstein, Am. J. Path. 68 (1972) 539.

(3) P.M. Henson, J. Immunol. 107 (1971) 1535.

(4) E.L. Becker, H.T. Showell, P.M. Henson and L.S. Hsu, J. Immunol. 12 (1974) 2047.

(5) C.G. Cochrane and B.S. Aikin, J. Exp. Med. 124 (1966) 733.

(6) A. Janoff, Am. J. Path. 68 (1972) 579.

(7) G.S. Lazarus, J.R. Daniels and J. Lian, Am. J. Path. 68 (1972) 565.

(8) A. Janoff and B.W. Zweifach, J. Exp. Med. 120 (1964) 747.

(9) E.S. Golub and J.K. Spitznagel, J. Immunol. 95 (1966) 1060.

(10) S.J. Klebanoff, Sem. Hematol. 12 (1975) 117.

(11) D.C. Smith and S.J. Klebanoff, Biol. Reprod. 3 (1970) 229.

(12) S.J. Klebanoff and D.C. Smith, Biol. Reprod. 3 (1970) 236.

(13) S.J. Klebanoff and R.A. Clark, Blood 45 (1975) 699.

(14) A. Böyum, Scand. J. Clin. Lab. Invest. Suppl. 21 (97) (1968) 77.

(15) R.A. Clark, S.J. Klebanoff, A.B. Einstein and A. Fefer, Blood 45 (1975) 161.

(16) P.J. Edelson and Z.A. Cohn, J. Exp. Med. 138 (1973) 318.

(17) H.J. Weiss, New Eng. J. Med., 293 (1975) 531, 580.

(18) R.T. Canoso, R. Rodvien, K. Scoon and P.H. Levine, Blood 43 (1974) 645.

(19) G.W. Philpott, R.J. Bower and C.W. Parker, Surgery 74 (1973) 51.

(20) J. Schultz, H. Snyder, N-C. Wu, N. Berger and M.J. Bonner, in: The Molecular Basis of Electron Transport, eds. J. Schultz and B.F. Cameron (Academic Press, New York, 1972) p. 301.

(21) H.I. Zeya and J.K. Spitznagel. J. Exp. Med. 127 (1968) 927.

(22) H. Odeberg and I. Olsson, J. Clin. Invest. 56 (1975) 1118.

(23) R.I. Lehrer, K.M. Landra and R.B. Hake, Infect. Immun. 11 (1975) 1226.

(24) R.A. Clark and S.J. Klebanoff, J. Exp. Med. 141 (1975) 1442.

(25) D.R. Phillips and M. Morrison, Biochemistry 10 (1971) 1766.

(26) A.L. Hubbard and Z.A. Cohn, J. Cell Biol. 55 (1972) 390.

(27) C-M. Tsai, C-C. Huang and E.S. Canellakis, Biochim. Biophys. Acta 332 (1973) 47.

(28) K. Agner, Proc. Int. Congr. Biochem. 4th Vienna 15 (1958) 64.

(29) A.U. Khan and M. Kasha, J. Chem. Phys. 39 (1963) 2105.

(30) R.C. Allen, Biochem. Biophys. Res. Commun. 63 (1975) 675.

(31) R.C. Allen, Biochem. Biophys. Res. Commun. 63 (1975) 684.

(32) M.M. King, E.K. Lai and P.B. McCay, J. Biol. Chem. 250 (1975) 6496.

Discussion

J. Schultz, Papanicolaou Cancer Research Institute: I enjoyed this paper very much and it follows the productivity of Dr. Klebanoff in this particular area. I am quite curious about the effect on the neighboring polymorphonuclear cells resulting from the chlorination reaction. Because you have such a consistent 25 or 26% kill could it be that, inasmuch as the populations of polymorphonuclear cells which is a dying population, you are selecting that portion of the population which would be particularly sensitive to that reaction while freshly produced polys would be more resistant. The results thus reflecting this distribution of the kinds of cells rather than an attack on all cells.

S. Klebanoff, University of Washington: You are talking now about the toxicity to the PMN?

J. Schultz: Yes

S. Klebanoff: We use Ficoll-Hypaque separated cells which are over 95% PMNs. The preparation contains small numbers of eosinophils and some mononuclear cells; however, it consists predominantly of neutrophils. In that population there are certainly some cells which are younger than others, but we have no information as to whether the age of the cells affects their susceptibility to the toxicity.

J. Schultz: Thank you.

A. Mildvan, Institute for Cancer Research: What was the source of the chemiluminescence in the myeloperoxidase deficient cells and can you determine, by an analysis of emission spectrum, what fraction of it is due to singlet oxygen and what might be due to other emissive components?

S. Klebanoff: Light emission by MPO-deficient leukocytes is lower than that of normal cells during the early post-phagocytic period. I cannot say what the source of the light emission by MPO-deficient cells is. The chemiluminescence by these cells is only slightly affected by azide and is almost entirely inhibited by superoxide dismutase suggesting that it is dependent on the superoxide anion. It has been suggested that the spontaneous dismutation of the superoxide anion generates singlet oxygen; however, recent studies by Milsson and Kearns (J. Phys. Chem. 78: 1681, 1974) and King et al., (J. Biol. Chem. 250: 6496, 1975) suggested that singlet oxygen may not be a product of the spontaneous dismutation of O_2^-. Other proposed mechanisms for the generation of a singlet oxygen from the superoxide anion include the interaction of O_2^- and OH. (Arneson, ARch. Biochem. Biophys. 136:352, 1970) or the interaction of O_2^- and H_2O_2 (Kellog and Fridovich, J. Biol. Chem. 250:8812, 1975). Light emission also may be due to the formation of an excited substrate other than singlet oxygen. With sensitive methods one should be able to measure the spectrum of the light from PMNs and determine whether the emission is that characteristic of singlet oxygen relaxation, but this has not been done.

J. Schultz: In all myeloperoxidase reports there should be a statement (although I am sure that all the results are due to myeloperoxidase) of the source and purity of the enzyme.

S. Klebanoff: The enzyme is obtained from several sources. MPO is purified by us from dog pyometrial pus by the method of Agner (Acta. Chem. Scand. 12:89, 1958). We also have used human MPO kindly supplied by you and Dr. I. Ilsson, Sweden. However, we generally use canine MPO prepared to very high purity by the Agner method.

R. Leif, Papanicolaou Cancer Research Institute: Returning to Dr. Schultz's uestion, you said that you Ficoll-Hypaque separated the cells. I presume the neutrophils sedimented to the bottom of the gradient, in which case they have been subjected to some degree of osmotic shock as well as the fact that Hypaque itself is a fairly toxic material. The reason this is not significant in the case of lymphocytes is they never see the Hypaque. Thus your separation procedure might also render these cells more susceptible to killing.

S. Klebanoff: We have in a number of studies of neutrophil function employed either a simple dextran-sedimented cell preparation or Ficoll-Hypaque cells and we cannot detect very much difference in their ability to carry out these functions in vitro. Certainly Ficoll-Hypaque separated cells are very active functionally at the end of this procedure, which is widely used due to the advantage of high purity. We lyse the red cells with either ammonium chloride, or more commonly by exposure to hypotonic saline for a brief period.

R. Estabrook, University of Texas: I have two short questions. Have you any evidence that suggests an alteration of the unsaturated fatty acids of the phospholipids of the membrane as a basis of cytotoxicity? Could a change in the phospholipids be the primary initiating step for lysis?

S. Klebanoff: We have not looked for evidence of lipid peroxidation. The formation of lipid peroxides by a mechanism involving the superoxide anion and singlet oxygen has been described and these agents generated under the conditions of our experiments may well initiate lipid peroxidation at the cell surface.

R. Estabrook: You did not extract your TCA precipitates with organic solvents?

S. Klebanoff: No.

R. Estabrook: Secondly, how did you quantitatively measure the concentration of superoxide?

S. Klebanoff: We used the superoxide dismutase inhibitable reduction of cytochrome C as a measure of superoxide.

CYTOTOXICITY OF THE SUPEROXIDE FREE RADICAL

JOE M. McCORD and MARVIN L. SALIN
Departments of Medicine and Biochemistry
Duke University Medical Center

Abstract: To determine the possible cytotoxic effects of the superoxide free radical, phagocytosing polymorphonuclear leukocytes, which liberate the radical, were compared to resting leukocytes with regard to maintenance of cell viability. Phagocytosing leukocytes died a premature death which could be prevented by the addition of superoxide dismutase, catalase, or mannitol to the medium. These data indicate that the actual cytotoxic species was the hydroxyl radical, OH·, generated secondarily via the reaction of superoxide with hydrogen peroxide. Superoxide production by phagocytes appears to be a deliberate offense mechanism which can be directed against other cell types, but which can also result in unwanted deleterious side-effects, damaging to the host.

INTRODUCTION

The superoxide free radical, O_2^-, is generated as a toxic intermediate in a wide variety of biological reactions that reduce molecular oxygen, and several specific mechanisms have been delineated by which its deleterious actions are realized (1,2). In 1968 an enzymic activity was discovered which catalyzes the dismutation of the superoxide radical to oxygen and hydrogen peroxide (3):

$$O_2^- + O_2^- + 2H^+ \rightarrow O_2 + H_2O_2 \ .$$

This enzyme, superoxide dismutase [EC 1.15.1.1], was thought to play a protective role to the organism by scavenging or detoxifying the radical (4), and a considerable amount of evidence has accumulated in support of this hypothesis (2,5-9).

Although it appears that all oxygen-metabolizing organisms contain large amounts of intracellular superoxide dismutase (SOD), the mammalian extracellular fluids that have been examined contain only traces of the activity (2).

An intriguing physiological role for superoxide was hypothesized by Babior *et al.* with the discovery that phagocytosing polymorphonuclear leukocytes (PMN), effector cells of the acute inflammatory response, release large amounts of superoxide into the medium in which the activated cells are suspended (10). Babior proposed that this normally noxious by-product of oxidative metabolism was deliberately produced by phagocytosing PMN as an offensive measure against invading microorganisms, and that the radical might therefore play a role in the bactericidal activity of such cells. This hypothesis was strongly supported by the studies of Yost and Fridovich (11) and by Johnston *et al.* (12). Since superoxide and secondary radicals are capable of killing ingested microorganisms, and since a considerable fraction of the total superoxide produced by the PMN appears to be produced on the outer surface of the cell membrane and released into the extracellular medium (13,14), we thought it likely that superoxide might display cytotoxic effects toward mammalian cell types. Perhaps the cell type most likely to be damaged or killed by exposure to the radicals is the very cell which produces the radical-- the activated phagocyte. We therefore sought to observe any self-inflicted, radical-mediated damage incurred by the PMN as a result of phagocytic stimulation.

EXPERIMENTAL

Human PMN were isolated from freshly drawn venous blood as previously described (13). The cells were typically greater than 95% viable as judged by their ability to exclude trypan blue (15) and were greater than 98% PMN. One ml incubation mixtures contained about 3×10^6 PMN suspended in Ringer-phosphate buffer, pH 7.4, containing 1% glucose and 5% autologous serum. When phagocytosing cells were desired, incubation mixtures contained *Escherichia coli* K-12 that had been washed in deionized water and stored frozen, at a ratio of ten bacteria per leukocyte. Other additions included, where indicated, bovine liver SOD, prepared as previously described (4), or bovine catalase (2× crystallized, Sigma Chemical Co., St. Louis, Mo.), bovine serum albumin (Sigma), or mannitol (Sigma). All incubation mixtures were maintained at 37°C in siliconized glass tubes. Longer incubations contained 50 U/ml of penicillin (Sigma) and 50 µg/ml streptomycin sulfate (Sigma) to retard extraneous bacterial growth. At the indicated times samples of the mixtures were withdrawn, mixed with a 1% solution of trypan blue, and examined microscopically. The percentage of cells not stained by the dye is represented as "% viable cells". In certain experiments a significant number of dead cells lysed. In these cases, the % viable cells was corrected for the decrease in total cell count.

The quantitative nitroblue tetrazolium (NBT) test was performed as previously described (14) in 10 × 75 mm siliconized test tubes, containing a total volume of 1.0 ml of Ringer-phosphate buffer with 5% autologous plasma. The incubation mixture contained 1.0 mM NBT and approximately 5×10^6 leukocytes. Where indicated, the leukocytes were stimulated by the addition of 10^8 *E. coli* cells which had been washed in deionized water and stored frozen. SOD was also added where indicated. Tubes were incubated in a water bath at 37°C and at intervals, aliquots were removed and 200 to 300 cells were microscopically counted. Cells containing the dark blue formazan particles were scored as NBT-positive. Results are expressed as percent NBT-positive cells. The endotoxin from *Salmonella typhosa* was obtained from Difco Laboratories, Detroit, Michigan. A lymphokine preparation was kindly provided by Mr. Linville Meadows.

RESULTS AND DISCUSSION

Fig. 1 shows that "resting" PMN or those incubated in the absence of *E. coli* maintained full ability to exclude trypan blue for approximately the first 18 hours of incubation. Thereafter, the percentage of viable cells slowly decreased, reaching a value of 70% after 35 hours. The cells incubated with *E. coli*, designated "phagocytosing", showed a slow loss of viability during the first 15 hours of incubation. After that time, loss of viability proceeded rapidly with 90% of the cells dead at 35 hours. A third incubation mixture of phagocytosing PMN contained *E. coli* plus SOD at 300 µg/ml. The addition of the SOD restored the viability of the phagocytosing cells to that of resting cells. After prolonged incubation, the pH of the mixtures dropped less than 1 unit, and this was not affected by the presence of protectants. Johnston and co-workers (12) found that SOD at comparable concentrations did not interfere with the process of phagocytosis and had little or no effect on bactericidal action by PMN.

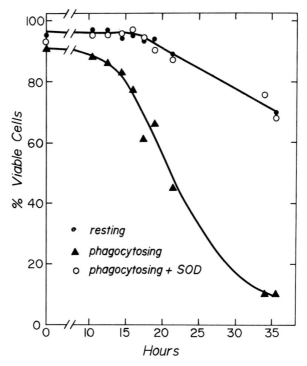

Fig. 1. Protection of phagocytosing PMN by SOD. Incubation mixtures contained 3×10^6 PMN in 1 ml of Ringer-phosphate buffer with 1% glucose and 5% serum. Those indicated "phagocytosing" contained 3×10^7 E. coli. SOD, where indicated, was at 300 μg/ml. At intervals samples were withdrawn, stained with trypan blue, and microscopically examined to determine percent viable cells.

The concentration of SOD required to produce maximal protection was determined by incubating phagocytosing suspensions of PMN with various concentrations of the enzyme at 37°C for 40 hours and then observing the viability of the cells. These data are shown in Fig. 2. The maximal protective effect was attained at 200 μg of SOD per ml. Most of the other phenomena which are inhibited by SOD have shown complete or maximal inhibition in the presence of much lower concentrations of SOD (see, e.g., ref. 2). One possible explanation was that, in this case, the generation of superoxide and its subsequent reactions were taking place on or very near the outer surface of the cell membrane. Since both the cell surface and the SOD are negatively charged at the pH of the reaction mixtures, it seemed possible that a charge repulsion

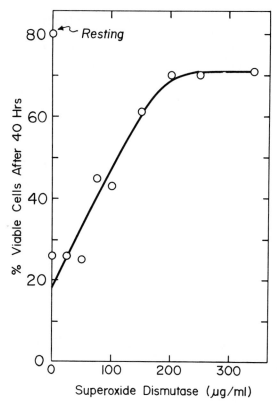

Fig. 2. Effect of SOD concentration on PMN viability. Conditions were as in Fig. 1, except SOD concentration was varied as indicated, and cells were examined after 40 hours. All cells were phagocytosing except the point labeled "Resting".

might partially exclude the SOD from the shell of solvent surrounding the cells in which the radical reactions were taking place. To test this hypothesis, we have prepared SOD which bears a net positive charge at pH 7.4 by covalently attaching polylysine (avg. mol. wt. 2000) to native SOD, using a carbodiimide as the cross-linking agent. In preliminary experiments, this derivatized SOD appears to be approximately ten times as effective as native SOD in protecting phagocytosing PMN under the conditions described in Fig. 1.

The nature of the protection afforded by SOD was further explored by incubation mixtures containing catalase or mannitol rather than SOD. In 1934, Haber and Weiss proposed the reaction: $O_2^- + H_2O_2 \rightarrow OH^- + OH\cdot + O_2^*$

by which the very powerful oxidizing radical OH· is produced
(16). Recently, Kellogg and Fridovich have produced evidence
that the oxygen produced as a product of this reaction is
singlet oxygen (O_2^*)(17). Several superoxide-dependent phenomena
have been shown to result from OH· produced secondarily
from superoxide by this reaction, including bactericidal
action by leukocytes (2,12,18). At least one such phenomenon,
the peroxidation of lipids, results from the singlet oxygen
produced secondarily via this reaction (17). Thus, it was a
possibility that the chemical species causing the loss of PMN
viability was either the hydroxyl radical or singlet oxygen,
rather than superoxide, *per se*. The production of OH· by the
Haber-Weiss reaction can be prevented by the action of SOD or
of catalase, either of which halts the reaction by scavenging
a reactant. Alternatively, the OH· may be scavenged as it is
formed by such agents as mannitol (19). Fig. 3 shows that
SOD, catalase, or mannitol maintained the viability of phagocytosing
PMN equal to that of resting cells. The failure of
bovine serum albumin to protect the cells rules out any nonspecific
protein effect, as does the fact that the incubation
mixtures contained 5% serum. We therefore conclude that the
agent causing the death of the cells is the hydroxyl radical
produced secondarily from the superoxide radical. Superoxide
radical, *per se*, which was not scavenged in those incubation
mixtures containing mannitol or catalase, did not cause the
premature death of the cell. Likewise, hydrogen peroxide,
per se, which was not scavenged in those incubation mixtures
containing mannitol or SOD, did not cause the premature death
of the cell. Only when O_2^- and H_2O_2 are simultaneously present
is hydroxyl radical formation possible, and only under these
conditions was premature death following phagocytosis observed.

The time-course of cell death varied considerably for
cells obtained from different donors, but the maximal rate of
cell death was always seen to take place between 12 and 30
hours after the phagocytic challenge. This delayed death was
not anticipated since the metabolic burst of O_2^- production
was shown to occur in the 30 min after phagocytosis. After
30 min the superoxide production of the cells returns to the
level of resting cells (10,14,20). Thus, it appeared that the
cells were "mortally wounded" by the initial burst of radical
production but that death did not result from this damage
until sometime later. To test this hypothesis, PMN were induced
to phagocytose in the presence and, absence of SOD.
After 1.5 hours, when superoxide production would have ceased,
samples of each incubation mixture were gently centrifuged
down and resuspended in media with or without SOD for the

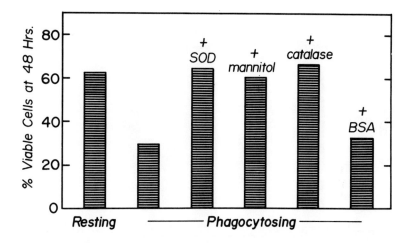

Fig. 3. Protection of phagocytosing cells by SOD, catalase, and mannitol. Incubation mixtures were as described in Fig. 1 with the following additions as indicated: SOD, 250 µg/ml; mannitol, 1 mM; catalase, 250 µg/ml; bovine serum albumin (BSA), 250 µg/ml. Viability was determined after 48 hours. These results represent the average of three separate experiments. Percent viability for unprotected cells was 32±8.5% (n=6), and for protected phagocytosing cells was 65±5% (n=9).

remainder of the observation period. The results of this experiment, seen in Fig. 4, clearly show that cell death can be prevented only if SOD is present during that part of the incubation in which the superoxide production takes place. SOD cannot prevent cell death if added after the exposure to superoxide has occurred.

Results identical to those described above were observed when PMN were stimulated with serum which had been incubated with bacteria, then centrifuged to remove the bacteria themselves before being added to the PMN. That is, the presence of particulate matter for inclusion into phagocytic vacuoles was not necessary for the observance of the premature death phenomenon or for its prevention by SOD. This finding rules out the possibility that the leukocyte death might be related to bacterial metabolic events taking place within the phago-

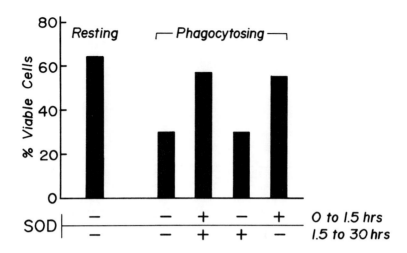

Fig. 4. Time-course of protection of phagocytosing PMN by SOD. Conditions were as described in Fig. 1. After allowing 1.5 h for the completion of phagocytic activity, the cells were centrifuged at 750 g for 5 min and resuspended in fresh medium, then incubated an additional 28.5 h. SOD was present, when and where indicated (+), at 300 µg/ml. The percentage of viable cells at the end of the incubation was based on the number of viable cells present at $t=0$. These data represent the results of a single experiment. The phenomenon was also observed with a 2 h initial incubation period.

cytic vacuoles. It is known that a variety of stimuli can cause increased rates of NBT reduction by PMN, and that particles to phagocytose are not an absolute requirement. Since there are two components to NBT reduction by PMN (a baseline rate not inhibited by SOD, and a stimulated rate which is totally due to superoxide liberated from the cells)(14), it seemed important to determine how each of these two components is affected by various other stimuli. Fig. 5 illustrates the effects of two soluble factors, endotoxin and lymphokine, on NBT reduction by isolated PMN. The endotoxin is a soluble lipopolysaccharide preparation from *S. typhosa*. Lymphokine represents one or more soluble mediators of inflammation released from lymphocytes which are incubated with a non-specific

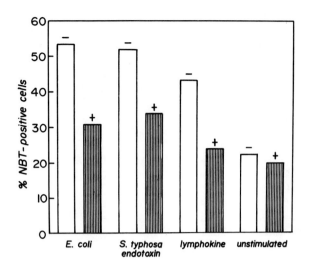

Fig. 5. Effects of several stimulants on superoxide production by PMN. Incubations were performed as described under EXPERIMENTAL, with the following additions: *E. coli*, 10^8 cells; *S. typhosa* endotoxin, 400 μg; lymphokine, a ten-fold dilution of supernate from 10^6 concanavalin A-stimulated lymphocytes per ml. SOD was present where indicated (+) at 450 μg/ml.

mitogen (such as concanavalin A) or with an antigen to which they have been sensitized (21). Lymphokine is a chemotactic factor for other leukocytes. While the baseline or uninhibitable NBT reduction showed a modest increase in the presence of all three stimuli, the superoxide mediated NBT reduction showed a dramatic increase in each case. We conclude that phagocytosis of particles is not the only circumstance which "turns on" superoxide production by PMN.

These studies have demonstrated a protective effect of SOD, catalase, and mannitol on the viability and integrity of phagocytosing PMN *in vitro*. This phenomenon of protection of cells against free radical death may contribute to the clinically observable anti-inflammatory activity of SOD (22,23). It is clear that the death of leukocytes and the concomitant or subsequent release of hydrolytic enzymes and chemotactic factors play major roles in the perpetuation of the inflammatory cycle (24). The cells of surrounding healthy tissue would, of course, also be subject to serious insult leading to death and lysis, either from primary attack by the radicals in the interstitial fluid or from secondary attack by hydrolytic enzymes released from lysed PMN. By intravenous injection, it may not be possible to maintain the necessary concentration of SOD in the extracellular fluids to protect PMN from their own free radical-induced destruction, since the protein is cleared quite rapidly from the circulation. By local injection, however, such a concentration may be realistically achieved at the site of the inflammation (22,23). As an approach to this problem, we have recently synthesized an enzymatically active derivative of SOD by covalently attaching polymers (avg. mol. wt. 2000) of polyethyleneglycol to ε-amino groups of lysine residues in native bovine SOD. When injected into rats, this derivative displays a dramatically increased circulating half-life (25). Attempts to assess the anti-inflammatory activity of this derivative *in vivo* are underway.

The data reported herein and elsewhere (2) support the notion that the superoxide radical *per se* may not be so detrimental a species in a physiological context. Through its reaction with H_2O_2, however, it is capable of generating another radical species, OH·, which is apparently capable of causing damage of grave consequence to physiological systems. Because of the nearly complete absence of SOD and catalase in extracellular fluids, the generation of OH· *in vivo* seems a real possibility. In the present case, the chemical nature of the damage to the PMN is unknown. The site of the lethal damage appears to have been the outer surface of the plasma membrane, since extracellular SOD and catalase prevented it.

The ability to produce superoxide and its use as an offensive chemical weapon do not appear to be unique to the bactericidal system of the PMN. We have shown that monocytes possess superoxide-producing capabilities (14), and the same has been shown for both alveolar and peritoneal macrophage (26). Furthermore, the stimulation of superoxide production by phagocytes has been demonstrated using complement and immunoglobulins (27). Although little is known at the present regarding the susceptibilities of cell types other than ingested microorganisms or the PMN itself to radical-induced damage, it is tempting to speculate that superoxide production may play a prominent role in certain types of autoimmune diseases, or in the eradication of random malignant mutations by the immune surveillance system.

REFERENCES

(1) I. Fridovich, Adv. Enzymol., 41 (1974) 35.

(2) J.M. McCord, Science, 185 (1974) 529.

(3) J.M. McCord and I. Fridovich, J. Biol. Chem., 243 (1968) 5753.

(4) J.M. McCord and I. Fridovich, J. Biol. Chem., 244 (1969) 6049.

(5) J.M. McCord, B.B. Keele, Jr., and I. Fridovich, Proc. Natl. Acad. Sci. U.S.A., 68 (1971) 1024.

(6) J.M. McCord, C.O. Beauchamp, S. Goscin, H.P. Misra, and I. Fridovich, in: Oxidases and Related Redox Systems, Vol. 1, eds. T.E. King, H.S. Mason and M. Morrison (University Park Press, Baltimore, Md., 1973) p. 51.

(7) F. Lavelle, A.M. Michelson and L. Dimitrijevic, Biochem. Biophys. Res. Commun., 55 (1973) 350.

(8) A.M. Michelson and M.E. Buckingham, Biochem. Biophys. Res. Commun., 58 (1974) 1079.

(9) A. Petkau and W.S. Chelack, Int. J. Radiat. Biol., 26 (1974) 421.

(10) B.M. Babior, R.S. Kipnes and J.T. Curnutte, J. Clin. Invest., 52 (1973) 741.

(11) F.J. Yost, Jr., and I. Fridovich, Arch. Biochem. Biophys., 161 (1974) 395.

(12) R.B. Johnston, Jr., B.B. Keele, Jr., H.P. Misra, L.S. Webb, J.E. Lehmeyer and K.V. Rajagopalan, in: The Phagocytic Cell in Host Resistance, eds. J.A. Bellanti and D.H. Dayton (Raven Press, New York, 1975) p. 61.

(13) M.L. Salin and J.M. McCord, J. Clin. Invest., 54 (1974) 1005.

(14) J.M. McCord and M.L. Salin, in: Erythrocyte Structure and Function, ed. G.J. Brewer (Alan R. Liss, Inc., New York, 1975) p. 731.

(15) H.J. Fallon, E. Frei, III, J.D. Davidson, J.S. Trier, and D. Burk, J. Lab. Clin. Med., 59 (1962) 779.

(16) F. Haber and J. Weiss, Proc. R. Soc. Ser. A, 147 (1934) 332.

(17) E.W. Kellogg, III, and I. Fridovich, J. Biol. Chem., 250 (1975) 8812.

(18) C.O. Beauchamp and I. Fridovich, J. Biol. Chem., 245 (1970) 4641.

(19) J.M. McCord and I. Fridovich, Photochem. Photobiol., 17 (1973) 115.

(20) J.T. Curnutte and B.M. Babior, J. Clin. Invest., 53 (1974) 1662.

(21) R. Snyderman and L. Altman, in: Annual Review of Allergy, ed. C.A. Frazier (Med. Exam. Publ. Co., Flushing, New York, 1973) p. 377.

(22) K. Lund-Olesen and K.B. Menander, Curr. Therap. Res., 16 (1974) 706.

(23) H. Marberger, G. Bartsch, W. Huber, K.B. Menander and T.L. Schulte, Curr. Therap. Res., 18 (1975) 466.

(24) J. Hirsch, in: The Inflammatory Process, eds. B.W. Zweifach, L. Grant and R.T. McClusky (Academic Press, Inc., New York, 1974), Vol. 1, p. 411.

(25) R.C. Dean and J.M. McCord, unpublished data.

(26) D.B. Drath and M.L. Karnovsky, J. Exptl. Med., 141 (1975) 257.

(27) I.M. Goldstein, D. Roos, H. Kaplan and G. Weissman, Clin. Res., 23 (1975) 304A.

Dr. Salin is supported by Postdoctoral Research Fellowship NS 01322 from the National Institute of Neurological Diseases and Stroke. This work was also supported by Research Grant AM 17091 from the National Institute of Arthritis, Metabolism, and Digestive Diseases.

Discussion

J. Harrison, Papanicolaou Cancer Research Institute: The monovalent reduction of oxygen to superoxide requires no protons, but for peroxide formation you need two. Is there any evidence or any precedent which suggests the nature of the reduced oxygen product from a flavoprotein oxidase could be pH dependent?

J. McCord, Duke University: Yes, there is. In the autoxidation of free reduced flavins (FMN, FAD, or riboflavin), as well as of reduced flavoproteins, the ratio of superoxide to hydrogen peroxide varies with pH.

J. Harrison: So the fact that you are getting superoxide from the cellular membrane, may not reflect the actual situation in phagocytic vacuole where the pH is somewhat lower.

J. McCord: That is true. The ratio of superoxide production to hydrogen peroxide production may be different in the phagocytic vacuole from that observed in the extracellular fluid.

R. Estabrook, University of Texas: Dr. McCord, I have a couple of question. You speak of hydroxyl radicals formed during the reaction. If it is indeed a hydroxyl radical of significant concentration that is accumulated, one should be able to detect this by electron paramagnetic resonance spectroscopy, rather than through the use of indirect measurements such as the quenching effect of mannitol. What happens when you do electron paramagnetic resonance studies?

J. McCord: I do not think the radical accumulates to any significant extent at all. It is an extremely reactive species and will, in fact, abstract a hydrogen from almost anything it encounters. I would expect that the steady state level of hydroxyl radical in all of these experiments is vanishingly small.

R. Estabrook: That being the case, then how far would you presume it could migrate from its site of generation in order to have a cytotoxic effect?

J. McCord: That is a difficult question to answer. I do not know. If, for example, it were sufficiently stable to migrate back into a cell, perhaps it could inflict damage on the DNA in the nucleus. It could then be potentially more damaging than if it reacts only with extracellular and surface components.

R. Estabrook: Another question I have concerns superoxide dismutase. What inhibitors are there of this enzyme other than cyanide?

J. McCord: Cyanide is the only good inhibitor, and it only inhibits the copper-zinc dismutases. It would obviously be of great experimental worth if other inhibitors were available. We have no inhibitors of the manganese- or iron-containing dismutases.

R. Estabrook: Lastly, could you clarify a point that I think has caused confusion and that is the activity of superoxide dismutase relative to the superoxide dismutase activity of the metal ion alone? What is the ratio of activities of coppper alone or copper plus manganese vs. superoxide dismutase of the same metal ion concentration. I hear that the protein confers only about a five-fold increase in activity; is this correct or incorrect?

J. McCord: The rate constant for the enzyme catalyzed dismutation is about 2×10^9 M^{-1} sec^{-1}. Thus, a consideration of rate constants gives the enzyme catalyzed reaction an advantage of 10^4 over the spontaneous reaction. Liver cell cytoplasm contains SOD at a concentration of about 2×10^{-5}M, whereas the steady state concentration of superoxide is probably less than 10^{-10}M. To dismute via SOD, a superoxide radical need only collide with an SOD molecule, an event more than 10^5 times as likely as colliding with another superoxide radical, as required for spontaneous dismutation. Thus, another advantage of at least 10^5 is gained by the enzyme catalyzed dismutation, for an overall advantage of more than 10^9 over the non-enyzme reaction.

R. Estabrook: Let us assume that there is no spontaneous dismutation but that a metal ion, such as copper or manganese, is present.

J. McCord: Aquo-copper, or copper that is not chelated at all, is an effective catalyst of superoxide dismutation. Copper chelated by EDTA has no such catalytic activity. What is important to understand is that free copper, or "naked" copper, does not exist in a physiological context. If one examines cell-free extracts of any source, the only superoxide dismuting activity one observes is due to the enzymes. In other words, all the non-SOD copper or non-SOD manganese in the cell is bound up in ways such that it cannot function as a superoxide dismutase.

M. Horowitz, New York Medical College: Have you any information on the effect of the ascorbic acid levels on the generation of either superoxide, hydrogen peroxide, or hydroxyl radicals in polymorphonucleocytes?

J. McCord: I have not examined ascorbic acid in this regard. There was a report (M. Nishikimi, Fed. Proc. $\underline{34}$, 624 (1975) abs 2298) that ascorbic acid is an effective scavenger of superoxide. The rate constant was reported to be 2.7×10^5 M^{-1}, sec^{-1} which is fairly good. I do not have any data on that subject.

THE FUNCTIONAL MECHANISM OF MYELOPEROXIDASE

John E. Harrison
Papanicolaou Cancer Research Institute
Miami, Florida 33123

INTRODUCTION

Myeloperoxidase, the peroxidase of the mammalian granulocyte, was first isolated by Agner (1) and subsequently crystallized by Agner (2) and by Schultz (3). The enzyme has partially characterized in terms of protein structure (4) and heme structure (5,6). Like the two other heme-containing mammalian peroxidases whose function has been studied in detail (thyroid and lactoperoxidases) myeloperoxidase is considered to function in the peroxidation of a halide ion, most notably, chloride ion. Agner reported that the addition of chloride ion to solutions containing myeloperoxidase and hydrogen peroxide lead to the formation of a species with the chemical properties of HOCl (hypochlorous acid)(7). Agner also gave a list of biological material which underwent oxidative transformation, or chlorination, when reaction mixtures were supplemented with chloride ions (8).

A number of workers have examined the bacteriocidal, fungicidal, and general cytotoxic properties of myeloperoxidase when supplemented with Cl^- (9,10,11). Chance (12) first reported that myeloperoxidase is converted sequentially into two peroxide compounds, the latter of which (compound II) has been described by Agner (13) and shown by Odajima to contain 1 oxidizing equivalent (per heme) with respect to the ferric peroxidase (14). The very unstable primary peroxide compound was found to form with a second order rate constant of $1.2 \& 10^7$ M^{-1} sec^{-1} (Chance, personal communication). This value is in agreement with that found by Morell and coworkers (15) using conventional kinetic methods.

Although it is now accepted that the preferred route of halide peroxidation is via a two electron process (with the known peroxidases), previous work on the mechanism of chloride peroxidation by myeloperoxidase lead Agner to suggest the involvement of compound II (16). Subsequently, Harrison and Schultz reported that chloride ion does not reduce compound II (17).

A second spectrophotometrically observable form, that formed upon the addition of chloride ion to myeloperoxidase at neutral or acid pH (16) was also proposed to play a role in the peroxidative mechanism (16,18). Conversely, Morrison and Bayse have noted that such complexes are formed with catalytically inactive proteins, and with anions which are not peroxidized (19). Stelmaszynska and Zgliczynski showed that the stability of the chloride complex of myeloperoxidase was directly dependent on the proton concentration (18).

A recent report from this laboratory has confirmed that myeloperoxidase catalyses the peroxidation of Cl^- to free HOCl (20). It was also noted that the addition of a chlorinable substrate had no detectable effect on turnover rate, implying that the rate-determining step in chloride peroxidation (under the conditions) was prior to the formation of a hypothetical chlorinating form of myeloperoxidase, similar to that described by Hager and coworkers for chloroperoxidase (21). It was also shown that unlike either chloroperoxidase or horseradish peroxidase, myeloperoxidase cannot utilize chlorite (ClO_2^-) as oxidant and chlorine atom source for chlorination.

The work described here is being performed to answer the following questions. Firstly, is compound I, or compound II (or both) a catalytic form in chloride peroxidation? Secondly, what is the role of the spectrophotometrically observed chloride complex and thirdly, how do the pH and Cl^- activity dependencies of chlorination equate with the second question and the biological function of the peroxidase.

Compound I is formed at the expense of both hydrogen peroxide and hypochlorous acid.

A partial answer to the first question has been obtained. Compound II, formed in the presence or absence of Cl^-, is not reduced by Cl^- at a significant rate (17). It has also been noted that compound II arises under conditions where the assays of chloride peroxidation (the chlorination of monochlorodimedone, followed at 280 nm (21,20)) exhibit marked non-linearity (rapid inactivation). This indicates that compound II is an inactive form of the peroxidase which does not function in the chloride peroxidase cycle.

Compound I can be observed to form as a precursor to compound II with both H_2O_2 and HOCl as oxidants (Fig. 1).

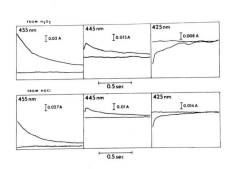
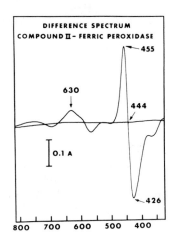

Figure 1 Left: Stopped flow traces recorded on mixing ferric myeloperoxidase with H_2O_2 and HOCl.

Right: Differences spectrum, myeloperoxidase compound II - myeloperoxidase.

The traces with H_2O_2 are in full agreement with those obtained by Chance (personal communication) using stopped-flow. Specifically, the change in absorbance at 455 nm, due to compound II formation (Fig. 1) exhibits a half time of around 200 msec under the conditions. This rate is too slow to account for the rate of turnover in Cl^- peroxidation. At 445 nm, the isosbestic point between compound II and the ferric peroxidase, the formation and decomposition of the primary peroxide compound is observed while at 425 nm, the biphasic increase in absorbance is due to the sequential (and simultaneous) formation of compound I and II. The spectral changes associated with the formation of the primary peroxide and hypochlorite compound are compatible with a broad diminution in the soret absorption, as observed for the primary compounds of other peroxidases (12). Further work is required to define the absorption spectrum of the primary compound. The relationship between the ferric peroxidase, compound I and compound II is shown in scheme I.

$$\text{Ferric MPO} \xrightarrow{H_2O_2} \text{compound I} \xleftarrow{HOCl} \text{Ferric MPO}$$
$$\downarrow$$
$$\text{compound II}$$

<u>Scheme 1</u>   Interpretation of stopped-flow data.

It should be pointed out that HOCl is a more strongly oxidizing under stopped flow conditions (high pH, very low Cl$^-$) than under assay conditions (low pH, high Cl$^-$), since oxidation of MPO by HOCl probably follows the equation: MPO + HOCl → $\underline{MPO_{OX}}$ + Cl$^-$ + H$^+$. Back oxidation of the peroxidase by product HOCl is probably insignificant under assay conditions, because of the high concentration of an alternative sink for HOCl and its low steady-state concentration. Under stopped-flow conditions, the rate of formation of compound I by HOCl is approximately an order of magnitude slower than that with an equivalent concentration of H$_2$O$_2$. Resolution of the formation of compound I, at the expense of H$_2$O$_2$ on a faster time scale than that shown above yielded a bimolecular rate constant of around 2×10^7 M^{-1} sec^{-1}.

The Chloride Complex

Fig. 2 shows a titration of ferric myeloperoxidase (canine) with chloride ion at pH 4.5. In a number of experiments, the average value for the apparent dissociation constant for Cl$^-$ was 0.3 mM at this pH.

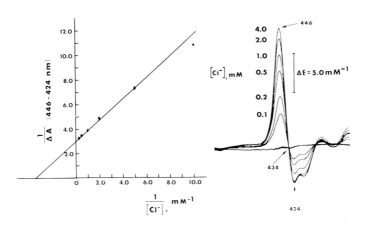

Figure 2 Determination of the apparent dissociation constant for chloride by spectrophotometric titration.

Chloride Inhibition of Peroxide Oxidation of Myeloperoxidase

When the rate of peroxide oxidation of myeloperoxidase (formation of compound I) is obtained by the method of Chance and Maehly (22) and chloride incorporated into the reaction mixture, spurious results arise because of the chlorination of, rather than oxidation of, the guaiacol donor. O-tolidine was therefore utilized at its maximum solubility; this donor gives the same product whether or not it is oxidized by the peroxidase directly, or by peroxidase-generated HOCl. The rate of product formation was obtained at varying chloride ion concentrations under conditions in which the peroxidation rate is determined primarily by the hydrogen peroxide concentration. Ideal "k1" conditions were not obtained, because of the limited solubility of o-tolidine, and the slow interaction of the donor at acid pH values. Fig. 3 shows representative traces; the rate of product formation, but not the yield of product, is lowered by increasing chloride concentration. By extrapolation of the rate to infinite chloride concentration (Fig. 3), the

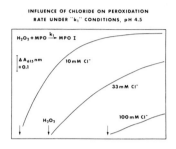
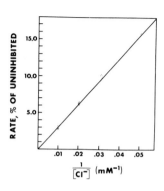

Figure 3 Left: Chloride inhibition of O-tolidine peroxidation under (predominantly) k_1 conditions.

Right: Reciprocal plot of rate data demonstrating complete inhibition at infinite chloride concentration.

peroxidation rate was found to be zero. The data are interpreted in terms of the blocking of peroxide oxidations of myeloperoxidase by chloride ion, the degree to inhibition being largely independent of the peroxide concentration utilized. To the extent that k_1 conditions pertain under the experimental conditions, the inhibition by Cl^- is compatible with the inactive form of the peroxidase being the spectrophotometrically observed chloride complex. This interpretation is shown in Scheme 2.

Chloride Complex $\xrightarrow{H^+, Cl^-}$ Ferric MPO $\xrightarrow{H_2O_2}$ Compound I

Scheme 2 Inhibition of "k_1" step by Cl^-.

Interpretation of pH, Cl⁻ activity curves in terms of chloride inhibition

The inhibition of peroxide oxidation by Cl^- provides a possible basis for the pH-activity profiles (constant chloride), and the effect of chloride concentration on these, as observed by Stelmaszynska and Zgliczynski (18). The interpretation utilizes the assumption that the bimolecular rate constant for the reduction of compound I by Cl^- is pH dependent.

Compound I + Cl^- + H^+ → Ferric Peroxidase + HOCl and v = k (compound I) $(Cl^-)(H^+)$

This represents an approximation of the pH dependence, based on the studies of Roman and Dunford (23) who showed that the rate of horseradish peroxidase compound I reduction by iodide was influenced by an ionization of HRP I outside the physiological range; a similar pH dependence for chloroperoxidase catalysed chlorination is observed (24). Furthermore, the work of Stelmaszynska and Zgliczynski, and work in this laboratory (Fig. 4) has demonstrated that catalytic activity increases with decreasing pH to an optimum. Since the interpretation of the positive dependence on H^+ can no longer be attributed to the pH-dependent stability of the spectrophotometrically-observed chloride complex, it can be concluded that the same pH dependence of anion reduction holds in myeloperoxidase as demonstrated in horseradish and chloroperoxidases.

A property of the chloride and pH dependence of chloride peroxidase activity not fully expounded by Stelmaszynska and Zgliczynski is inhibition by chloride (Fig. 4). A proposed basis for the occurrence of optima in both pH and (Cl^-) dependences is shown in Scheme 3. Chloride participates in the catalytic reaction in two ways (together with a proton).

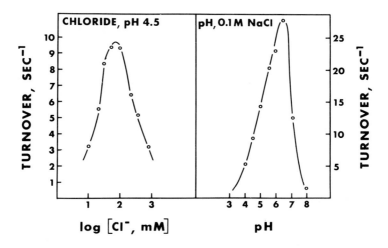

Figure 4 Influence of pH (at constant chloride concentration) and chloride ion concentration (at constant pH) on the rate of chloride peroxidation.

$$\text{CHLORIDE COMPLEX} \underset{K}{\overset{H^+, Cl^-}{\rightleftharpoons}} \text{MPO} \overset{H_2O_2, k_1}{\underset{k_2}{\rightleftharpoons}} \text{COMPD. I}$$
$$\text{HOCl} \qquad H^+, Cl^-$$

Scheme 3 Dual participation of Cl^- (and H^+) in chloride and pH dependence.

Firstly, in the product forming step, it reduces compound I, and secondly it complexes with the ferric peroxidase (substrate inhibition). The rate equation for Scheme 1 is given by:

$$\frac{v}{Et} = \frac{1}{\dfrac{K_h}{[H_2O_2]k_2K_D} + \dfrac{[H^+][Cl^-]}{k_1[H_2O_2]K_D} + \dfrac{K_h}{[H_2O_2]k_2[H^+][Cl^-]} + \dfrac{1}{k_1[H_2O_2]} + \dfrac{1}{k_2[H^+][Cl^-]}}$$

where K_h is the dissociation constant for H_2O_2.

Conditions under which optima occur in V/E versus H^+ or V/E versus (Cl^-) profiles can be obtained from this equation by differentiation. It is then found that the Cl^- optima are dependent upon pH and on the peroxide concentration.

Unfortunately, a major problem associated with the testing of the rate equation over a wide range of conditions is the poor reliability of data obtained under 'k_2' conditions. Thus under conditions of low (Cl^-) or high pH, (a) inactive compound II is formed rapidly, and (b) the kinetic assay (monochlorodimedone chlorination) exhibits marked nonlinearity. (A stopped-flow kinetic method is required to obtain turnover rates at very early times). We have attempted, however to fit equation 1 to a single set of data, using the experimentally determined constants of k_1 (2×10^7 M^{-1} sec^{-1}) and $K_D^{Cl^-}$ (0.3×10^{-3} M). Fig. 5 shows the result, for $k_2 = 10^3$ M^{-1} sec^{-1} (pH 4.5) and k_h (for H_2O_2) of 10^{-6} M.

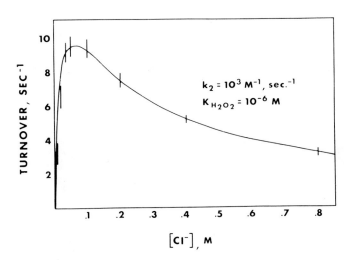

Figure 5

Fit of a single set of data to equation 1, for $k_2 = 10^3$ M^{-1} sec^{-1} and $K_{H_2O_2} = 10^{-6} M$ (fitted) and $K_{Cl^-} = 0.3$ mM and $k_1 = 2 \times 10^7$ M^{-1} sec^{-1} (experimental).

Relevance to the Biological Function of Myeloperoxidase

It appears that the kinetic properties, and the stabilities of the higher oxidation states of myeloperoxidase function to maintain strict control of the generation of HOCl. While refinement of kinetic data requires assays over shorter time periods, it is clear from the present work that a low pH (a) increases the rate of Cl⁻ peroxidation and (b) as a result of (a), increases the kinetic stability of the system. Which of the two related effects are of more importance in vivo requires experimental data on the steady state concentration of MPO compound I during the phagocytic process. It is also clear that the product forming step is measurable reversible, as evidenced by the fact that compound I arises from HOCl. This might imply that myeloperoxidase functions in the phagocytic process until the organic sink for generated HOCl is exhausted; at this point, myeloperoxidase would equilibrate with its product and rapidly convert to compound II. Experiments have shown that compound II arises from HOCl at acid pH (17) in the absence of Cl⁻, and at neutral pH with or without Cl⁻. There are therefore numerous factors which may function in the shutting down of chloride peroxidase activity.

Although several of the properties of the myeloperoxidase-catalysed chloride peroxidation can be rationalized in terms of the in vivo function, the property of substrate inhibition by Cl⁻ remains an enigma. Further work on MPO in situ (25) is required to determine to what extent the Cl⁻ complex is formed.

Acknowledgement

Supported by National Institute of Health grant No. R01-CA 10904.

1. Agner Kj. Acta Physiol. Scand. 2 (1941) Suppl. VIII.

2. Agner Kj. Acta Chem. Scand. 12 (1958) 89.

3. Schultz, J. Ann N.Y. Acad. Sci. 75 (1958) 22.

4. Schultz, J. Snyder, H., Wu, N.C., Berger, N. and Bonner, M.J. in The Molecular Basis of Electron Transport (J. Schultz and B.F. Cameron, editors) Academic Press (1972) 301.

5. Nichol, A.W., Morell, D.B., and Thompson, J. Biochem. Biophys. Res. Commun. 36 (1969) 576.

6. Wu, N.C. and Schultz, J. Febs Lett. 60 (1975) 141.

7. Agner Kj. Abstracts 4th International Congress of Biochemistry (1958) 5-58.

8. Agner, Kj. J. Exptl. Med. 92 (1950) 337.

9. Klebanoff, S.J. Ann. Rev. Med. 42 (1971) 39.

10. Klebanoff, S.J. In The Molecular Basis of Electron Transport (J. Schultz, and B.F. Cameron, editors) Academic Press (1972) 275.

11. Sbarra, A.J., Jacobs, A.A., Stauss, R.R., Paul, B.B. and Mitchell, G.W. Am. J. Clin. Nutr. 24 (1971) 272.

12. Chance B. in The Enzymes (J.B. Summer and K. Myrback, editors) Vol II, Part I, Academic Press (1951) 428.

13. Agner, Kj. Acta. Chem. Scand. 17 Suppl 1 (1963) 332.

14. Odajima, T and Yamazaki, I. Biochem. Biophys. Acta. 206 (1970) 71.

15. Newton, N., Morrell, D.B., and Clarke, L. Biochem. Biophys. Acta 96 (1965) 463.

16. Agner, Kj. in The Structure and Function of Oxidation-Reduction Enzymes; Proceedings of the Werner-Gren Symposium, Stockholm, Pergamon Press (1970) 329.

17. Harrison, J.E. and Schultz, J. American Chemical Society Abstract (1974) 99.

18. Stelmasnynoka, T. and Zglicmynski, J.M. Eur. J. Biochem. 45 (1974) 305.

19. Morrison, M. and Bayse, G. in <u>Oxidases and Related Redox Systems</u> (T.E. King, H.S. Mason and M. Morrison, editors) University Park Press (1973) 311.

20. Harrison, J.E. and Schultz, J. J. Biol. Chem. 251 (1976) 1371.

21. Thomas, J.A., Morris, D.R. and Hager, L.P. J. Biol. Chem. 245 (1970) 3129.

22. Chance, B. and Maehly, A.C. in <u>Methods in Enzymology</u>, Vol II (S.P. Colowick and N.O. Kaplan, editors) Academic Press (1955) 764.

23. Roman, R. and Dunford, H.B. Biochemistry II (1972) 2076.

24. Hager, L.P., Thomas, J.A. and Morris, D.R. in <u>Biochemistry of the Phagocytic Process</u> (J. Schultz, editor) North-Holland Publishing Company (1970) 67.

25. Wever, R., Roos, D., Weening, R.S., Vulsma, T. and Van Gelder, B.F. Biochem. Biophys. Acta 421 (1976) 328.

Discussion

S. Weinhouse, Temple University: How specific is the myleoperoxidase, or any of the peroxidases for that matter, to inorganic chloride? Will they accept chlorine from organic chlorides?

J. Harrison, Papanicolaou Cancer Research Institute: I am really not sure whether removal of chloride from organic substrates has been reported. I would expect that it can, but that only myeloperoxidase or chloroperoxidase would do it. Of course iodide has been reported to be available from the thyroid hormones.

S. Weinhouse: The reason I ask is that there are enzymatic mechanisms for dechlorination of organic chlorides of all sorts, and very little is known about them.

S. Klebanoff, University of Washington: Following the early studies with myeloperoxidase by Agner, it was proposed that the molecule contains two unequal hemes and that a possible mechanism for the peroxidation of chloride involves the formation of a complex between chloride and one of the hemes and a complex between hydrogen peroxide and the other heme and that an interaction between these two complexes could possibly explain the relatively unique ability of myeloperoxidase to catalyze the peroxidation of chloride. Do you have any comment on that possibility?

J. Harrison: Well, first of all I personally am convinced that the subunits of myeloperoxidase are functionally identical. Secondly, the inhibition of peroxidation by chloride parallels quite well the extent of chloride complex formation which one observes spectrophotometrically. Therefore, it does not seem that a half-saturated molecule could be the active form. In other words the substrate inhibition curve shows the same Km as the spectrophotometric titration with chloride.

MYELOPEROXIDASE-ENZYME THERAPY ON RAT MAMMARY TUMORS

J. SCHULTZ, A. BAKER*, B. TUCKER
Papanicolaou Cancer Research Institute
Miami, Florida 33123

Myeloperoxidase Enzyme Therapy On Rat Mammary Tumor

First I wish to pay tribute to Dr. Hugo Theorell in whose laboratory myeloperoxidase (MPO) was first discovered in 1941 by his student Kjel Agner (1). Dr. Agner's thesis has served as a Bible to those of us who have been working with myeloperoxidase over the last twenty-five years and one can rest assured that it will be some time before another source can replace it. This is true, although a greal deal of new information has been forthcoming since that time. Dr. Agner naturally thought about what myeloperoxidase was doing in the cell; he did carry out a series of experiments in which he demonstrated that the peroxidase can detoxify diptheria toxin and it gave him a hint as to its possible function.

Then along came the discovery of children's granulomatous disease, in which the myeloperoxidase is unavailable following phagocytosis (2). Such a child suffers from chronic infection. Then Klebanoff demonstrated that myeloperoxidase, halide, and hydrogen peroxide had a strong cytotoxic effect against bacteria (3), and Cline and Lehrer (4) found that there were patients with chronic candida infections who had no myeloperoxidase in the white cell.

It became clearly evident that the function of myeloperoxidase had to do with the defense mechanisms involved in the phagocytic process. Most of us who worked in the field before this recent knowledge became available now realize that we worked on a disease before it was discovered.

The intense interest in the killing of bacteria, virus and fungi by the MPO system lead to the recent attention being given the cytotoxicity of the myeloperoxidase system (5). That involving tumor cells aroused our interest be-

*Present address: Smith, Kline & French Laboratories, 1500 Spring Garden, Philadelphia, Pa.

cause of our studies on mammary adenocarcinoma induced by the methylcholanthrene a few years ago. We were interested in that phenomenon in the intact rat (6).

I think I owe it to my colleagues who are sophisticated biochemists, to explain why I should be involved in feeding rats, weighing tumors, measuring tumors, and restricting the experiment to that type of a technique. This did come about and I would like to explain it.

Dr. Harry Shay at the Fels Institute was trying to develop a gastric cancer by installation of methylcholanthrene directly into the rat's stomach. And to his surprise, the rats developed mammary adenocarcinoma (7). In addition to that, a few rats out of a thousand developed a chloroleukemia, which at that time was only known through certain cases of myelogenous leukemia in man where green tumors appeared in the sternum. It so happens that this particular chloroleukemia when injected subcutaneously into rat pups developed into large 50 to 80 gm tumors as the rats reached adulthood (8). Previous to this, in 1941, Dr. Agner being aware of this green tumor in man and of the search for the origin of the green color, examined a 17-year-old tumor from a museum jar in which the tumor was preserved in formaldehyde. He was able to speculate from spectrophotometric data that myeloperoxidase was responsible for the green color of the tumor (1). When I joined the Fels Institute staff, the green tumor aroused my curiosity; from it myeloperoxidase was isolated and crystallized (9,10).

The Lederle Laboratories used Dr. Harry Shay's chloroleukemic rat to develop thiotepa which cures that disease in the rat and is used against myelogenous leukemia in man. Of particular importance was that thiotepa was ineffective against lymphatic leukemia. One of the striking differences between the lymphocyte and the myelocyte is that myelocyte is rich in myeloperoxidase, up to 2-4% dry weight. The lymphocyte has no peroxidase. This was dramatically demonstrated when rats bearing 50-60 gram chloroma tumors were injected with thiotepa, and the tumors disappeared in five to seven days (11). Mammary tumors under the same conditions were resistant. Would such tumors respond if the constituents of chloroma could be incorporated in the tumor?

There was no intention to initiate a chemotherapeutic program as such, but rather to test the hypothesis that the specificity of thiotepa for myelogenous leukemia and/or chloroma was due to factors present in the tumor or the granulocyte that participated in the cell destruction by the alkylating agent. In an earlier report (12) hematoporphyrin was tested, because besides peroxidase, the chloroma tumors sensitive to the drug are rich in porphyrins (13). It is possible that the effect of the porphyrin may have been related to the peroxidase activity in their combination with cellular protein and iron.

The extremely limited amounts of the enzyme that was and is available made extensive studies impossible. The large numbers of control MCA treated animals (over 400) used in these experiments were the results of an attempt to find chloroleukemic rats as a source of myeloperoxidase. Although biweekly blood counts were taken on each animal throughout the course of MCA administration up to tumor formation, no leukemic animals were found over a period of several years. This has been the experience of other laboratories.

Included in this report are the experiments conducted on the fate of injected myeloperoxidase. The results served as a basis for injecting thiotepa 48 hours after the enzyme. Subsequent experiments, (some years later) after the procedure for determining the enzyme content of the white cells of a single sample of tail blood could be designed, it was learned that the enzyme concentrated in the PMN cell where it reached a maximum level six hours after injection. It might be added in this connection that when all the enzyme found in the tissues of the normal rat was summed up it represented less than 20% of the amount injected. The remainder was thought to be found in the PMN cells. Because of this, in the latest experiments thiotepa was injected six hours after the enzyme. They are included separately in Fig. 6 (curve 3), and can be seen to be quite impressive.

Methods and Results

The animals used in these experiments were treated essentially by those described by Shay and subsequently Shimkin (14), that is, 10 mg. methylcholanthrene (MCA) in olive oil on Mondays, Wednesdays and Fridays. Shimkin and co-workers described in great detail the nature of the tumors developed; the present authors made no effort to enter into the pathologic problems involved. These are reviewed by Foulds in his second volume of Neoplastic Development (15). However, the Shimkin group did carry out extensive chemotherapeutic trials in which the measurement of tumor volume over a period of 21 to 42 days was used to compare the effectiveness of various anti-neoplastic drugs (16,17,18,19). Similar parameters are used in the present report and it can readily be seen that there is general agreement that the tumor size just about doubles in the 21 to 42 days. We did not see the spontaneous regressions observed by others (15).

Sprague-Dawley female rats of about 140-150 gm were given 20-methylcholanthrene in sesame oil three times a week at the level of 10 mg per rat by stomach tube. Within 60-90 days tumor appeared; as each tumor reached about 3 cm^2 in size, as the product of two dimensions, they were put on experiment.

The first preparations of enzyme used were obtained from rat chloroma tumors by procedures described by us elsewhere (20). In Fig. 1 the results are described of a trial series where the enzyme level was 20 mg/Kilo and thiotepa 5 mg/Kilo given 48 hours after the enzyme.

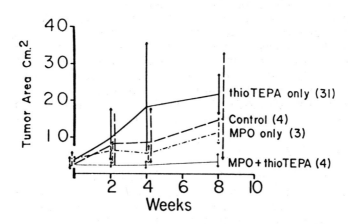

FIGURE 1

It can be seen there that the MPO and thiotepa effectively restricted the tumor growth during the period of the experiment, and that this group showed little of the variation in growth seen in the controls (saline injected), the thiotepa or the MPO treated groups.

All subsequent experiments were carried out with the MPO prepared from normal human white cells (21). The dosage levels of the enzyme were 24 mg/kilo, or 48 mg/kilo; with repeated doses no acute toxicity was noticed. This is remarkable for one must consider 48 mg/kilo a rather high level; and since the enzyme was prepared from another species, some foreign protein reaction was expected if any non-enzyme protein were present. Examples of the growth of individual tumors under various conditions of treatment and doses are shown. (Figs. 2 and 3). Tumor size indicated on the ordinate refers to changes in size since day of injection at 0 time.

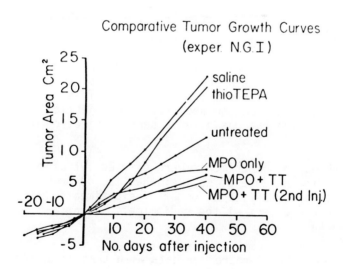

FIGURE 2

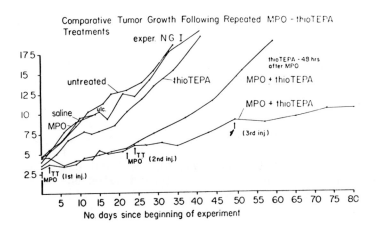

FIGURE 3

In Fig. 2 repeated injection of enzyme showed no injurious effect that might result from anaphylaxis. In a second such experiment the rate of tumor growth was controlled. This involved selecting those animals whose tumors were growing at approximately the same rate. This is seen in the -20 to 0 days on the abcissa. The injection took place at time 0. One rat was given a second injection on the 25th day.

Subsequent experiments were carried out at twofold increase in the dose of enzyme; that is, 48 mg MPO/kilo. Fig. 4 is a summary of a number of experiments. Here again the combined effect at the radiomimetic agent and the enzyme is remarkable.

In order to compare our data with that of Shimkin and co-workers (16) who plotted tumor volume at the 21st and 42nd day, we have combined all of our data and made a similar plot. Although the ordinate in the present data is the product of two dimensions and in Shimkin data the abcissa is the $\log_{10}$ of the volume, the course of growth is quite the same; thus, in the untreated controls (415 rats) there is a doubling in the tumor size (curve 9 of Fig. 5).

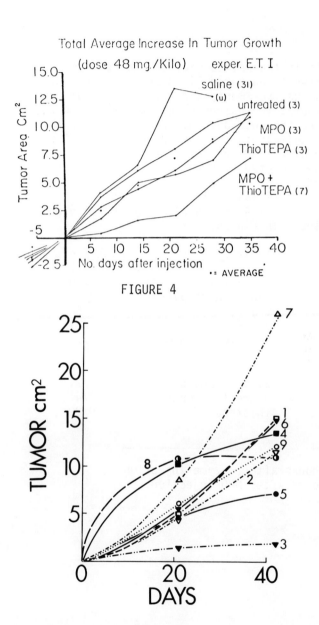

FIGURE 4

FIGURE 5

TABLE I

This table is the key to Fig. 5, the numbers in the first column refer to numbered lines on the chart. The second column is the treatment; and the third column represents the survival data, the numerator being the number of rats that died before the 40th day and the denominator the initial number of animals in that group; and the last column this is expressed as %. MPO refers to human enzyme, VPO to the peroxidase isolated from rat chloroma. This is arbitrary, but at the time of preparation from chloroma, the peroxidase of the leucocyte was called Verdoperoxidase. Thiotepa treated (*) was 7.4 on the 21st day and 16.5 on the 42nd day (not included in Fig. 5).

Curve No.	Treatment	Deaths/Initial No.	%
1.	MPO	1/21	4.8
2.	MPO + Thiotepa(**)	4/30	13.
3.	MPO + Thiotepa(+)	0/3	0
4.	VPO	1/5	20.
5.	VPO + Thiotepa(**)	3/11	27.
6.	Inactivated MPO	0/7	0
7.	Inactivated MPO + Thiotepa(**)	1/7	14.
8.	Saline Controls	11/21	52.
9.	Untreated Controls	100/415	24.
(*)	Thiotepa	9/29	31.

(**) 48 hours after enzyme injection
(+) 6 hours after MPO Injection.

The curves show that combining all rats in one group can lead to errors, because the data shown in Figs. 1-5 a single injection on day 0 would restrict growth for 10-15 days and by the end of the 20 days the effect of MPO and thiotepa (curves 2,3,5) when compared to other treatments (curves 4,7,8) such as saline, inactive enzyme with or without thiotepa (curves 8,4,6,7), are quite distinct by day 20. The effect of deaths on the shape of the curve from the 20th to the 40th day can confuse the issue because the rats with the most rapidly growing tumors may have died leaving the slower growing tumors to give the impression of slower overall growth. For this reason mortality data is included. One might note that considering only those groups with 20 or more rats, the mortality of the enzyme treated animals (5-13%) are quite striking compared to the controls or treated rats (24-52%).

Thiotepa treated rats are not shown in Fig. 5, but the figure for the 21st day was 7.4 and that for the 42nd day was 16.5 placing it above all the rest, except curve 7.

One can only conclude that the regimen of MPO and thiotepa may be as effective as the better drugs of the numerous antineoplastic agents examined by previous investigators using similar criteria (16-19). The explanation for this is the subject of current researches.

Fate of MPO Following I.P. Injection

The purpose of the following experiments was to determine the distribution in the tissues of intraperitoneally injected enzyme. Since the myeloperoxidase was made up in saline, a series of saline injected controls were used in each experiment. Two levels of enzyme, 24, and 48 mgs per kilo gm body weight, were injected intraperitoneally to a series of 200-210 gm rats. Normal animals were sacrificed on time zero, and at subsequent time intervals up to 72 hours. When thiotepa was used, it was injected 48 hours after the myeloperoxidase. At time of sacrifice, liver, lungs, kidneys and spleen were removed and homogenized in 0.1 M phosphate buffer pH 7.0 at 1 gm per 100 cc buffer. Aliquots were frozen and thawed 6 times in presence of triton for the enzyme assay and another aliquot taken for protein analysis by the Lowry procedure.

In the normal rat only the lungs and spleen showed peroxidase activity. Following saline injection there is depression at about 8 hours which value returns to normal in 24 hours. (See Fig. 6, curves marked "c" are <u>saline controls.</u>) In the same figure the injection of 4,800 K of enzyme per 100 gm body weight results in a sharp rise in the spleen within 4 hours, followed by increases in the lung and liver. These experiments on non-tumor bearing rats illustrate that in normal animals the enzyme distribution returned to base line values in 24 hours.

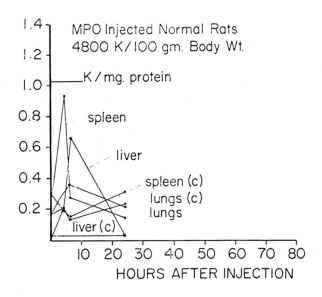

FIGURE 6

In subsequent experiments only tumor bearing rats were used and the controls were saline injected animals of the same size and age of tumor as was possible. The time period was also extended as indicated by the longer time required to return to normal. In the figure illustrating the results, those curves marked "c" are the saline controls. An example of the effect of saline is shown in Fig. 7 where the tumor is seen to show little or no peroxidase activity for 72 hours, while the lungs and spleen rise slightly and remain

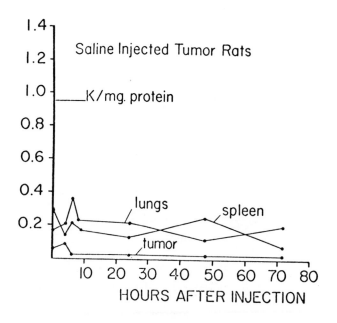

FIGURE 7

constant for 72 hours. The liver and kidney of the tumor rat show no activity measurable under these conditions. The effect of thiotepa by itself when compared to saline controls in the tumor bearing rat is seen in Fig. 8. Sharp initial responses are noted within 4 hours in the case of the lung and spleen. Within 48 hours the tumor appears to acquire some enzyme as if the loss from the lungs and spleen resulted in a migration to the tumor. This can be observed more remarkably after injection of 24 mg/Kilo into tumor bearing rats where a tremendous accumulation takes place in the spleen and as it diminishes, it appears in the tumor at 72 hours (Fig. 9). A similar change appears at 4 and 6 hours. When repeated with 48 mg of enzyme per Kilogram body weight, the amount in the spleen almost doubles. Quite different from the normal rat at this level of enzyme, the liver and kidneys show marked accumulations; much higher amounts of enzyme are also found in the lung. Again as with the 24 mg injection, there is an increase deposit of MPO in the tumor at 72 hours.

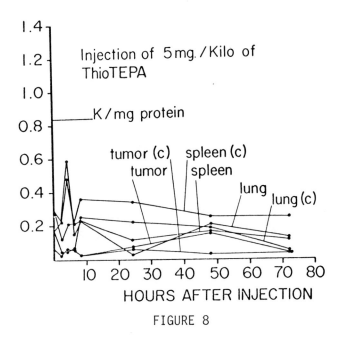

FIGURE 8

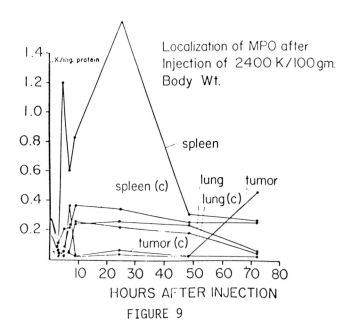

FIGURE 9

The effect of thiotepa superimposed on the injection of myeloperoxidase resulted in an earlier and more prolonged elevation above baseline levels as seen in Fig. 10. Since

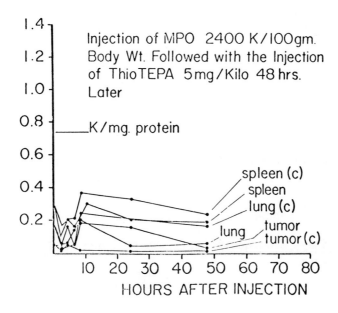

FIGURE 10

thiotepa was injected 48 hours after the myeloperoxidase, zero time represents in this case 47 hours after the myeloperoxidase when compared to Fig. 8. Thus within 4 to 8 hours the MPO begins to accumulate in the tumor and remains above control values up to 96 hours. The saline controls in the case of the lung and spleen have more enzyme than the experimental rats originally injected with MPO and thiotepa, but this is not true of the tumor itself. It was on the basis of these data that thiotepa was injected 48 hrs. after the enzyme. But subsequent experiments wherein the white cell were examined following MPO injection, the enzyme was found in the neutrophile (22) reaching a maximum in 6 hours. A few experiments using MPO followed by thiotepa in 6 hrs. were found to be rather effective (Fig. 5).

Bibliography

1. K. Agner, Acta Physiol. Scand. 2 (1941)Suppl. VII.
2. B. Holmes, A.R. Page and R.A. Good, J. Clin. Invest. 46(1967)1422.
3. S.J. Klebanoff, Ann. Rev. Med. 22(1971)39.
4. R.I. Lehrer and M.J. Cline, J. Clin. Invest. 48(1969) 1478.
5. S.J. Klebanoff (This Volume).
6. J. Schultz, A. Baker, B. Tucker, Proc. Amer. Assoc. Can. Res. 5(1964)57.
7. H. Shay, E.A. Aegerter, M. Gruenstein and S.A. Kamorov, J. Nat. Cancer Inst. 10(1949)255.
8. H. Shay, M. Gruenstein, C. Harris, and L. Glazer, Blood 7(1952)613.
9. J. Schultz, 131st Meeting Amer. Chem. Soc. (1957) 12C.
10. J. Schultz, Annals N.Y. Acad. Sci. 75(1958)22.
11. J. Schultz, A. Turtle, H. Shay, M. Gruenstein, 128th Meeting Amer. Chem. Soc. (1955) 70C.
12. J. Schultz, Union Inter. Contre Cancer 16(1960)769.
13. J. Schultz, H. Shay, M. Gruenstein, Cancer Research 14(1954)157 and J. Schultz and S. Schwartz, Ibid, 16(1956)565.
14. H. Shay, M. Gruenstein, W.B. Kessler, J. Natl. Cancer Inst. 27(1961)503.
15. L. Foulds, Neoplastic Development, Vol 2 (Academic Press, N.Y., 1975) p. 509ff.
16. L. Gropper and M.B. Shimkin, Cancer Res. 27(1967)26.

17. A.P. Davis, L. Gropper and A.B. Shimkin, Cancer Res. 26 (part 2) (1966)19.

18. A.P. DAvis, M. Gruenstein and M.B. Shimkin, Cancer Res. 26(part 2) (1966)1.

19. M.B. Shimkin, L. Gropper and M. Gruenstein, Cancer Res. 27 (part 2) (1967)1284.

20. J. Schultz, A. Gordon, and H. Shay, J. Amer.Chem. Soc. 79(1957)1632.

21. J. Schultz and H. Shmukler, Biochemistry 3(1964)1234.

22. J. Schultz, H. Snyder, N.C. Wu, N. Berger and M.J. Bonner, The Molecular Basis of Electron Transport (Academic Press, N.Y., 1972) p. 301.

Acknowledgment

These researches were supported in part by NIH grants CA-3715, CY-5152 and CA-10904. We also wish to express our appreciation for the technical assistance of Nancy Green and Marcia Specter.

Discussion

A. Mildvan, *Institute for Cancer Research*: Does hydrogen peroxide or organic peroxide have any effect on these tumors?

J. Schultz, *Papanicolaou Cancer Research Institute*: Well, there are some experiments done at NIH in which they put tumor-bearing rats in hyperbaric oxygen and they did see some effect. For a long time they had experiments of that type going on and it is quite possible that hydrogen peroxide developed in those experiments stimulated the leukocyte to become cytotoxic.

R. Leif, *Papanicolaou Cancer Research Institute*: The finding that the myeloperoxidase is concentrated by the neutrophils has a very interesting implication for chronic granulomatous disease. Has anyone tried feeding human myeloperoxidase, which would not have the side effects of blood transfusions, to these people.

J. Schultz: Not that I know of Dr. Leif. You know it is a question of discouraging use of myeloperoxidase as a possible therapeutic agent since it costs $7 million a pound to make. However, there are some porphyrins which we did try years ago in the days when Dr. Sam Schwartz was excited about injecting porphyrins into tumor-bearing rats and studying their increased sensitivity to radiation. Dr. Schwartz is very actively pursuing that but I always thought that the porphyrins going into the cell would somehow or other, like hemes, develop peroxidase activity. That is one reason, amongst others, that we changed to myeloperoxidase. However, it is not the myeloperoxidase that is deficient in childrens granulomatous disease. It is the lack of H_2O_2 generators that is missing; for example NADH or NADPH oxidase.

DEFECTIVE REGULATION OF CHOLESTEROL BIOSYNTHESIS IN TUMOR-
VIRUS TRANSFORMED AND HYPERCHOLESTEROLEMIC HUMAN SKIN FIBRO-
BLASTS: A COMPARATIVE STUDY

J. Martyn Bailey, Todd Allan, E. Jean Butler and Jiunn-Der
Wu. Department of Biochemistry, School of Medicine, George
Washington University, Washington, D. C.

Human fibroblasts from normal subjects, homozygous
Type II hypercholesterolemic individuals and SV40 tumor-
virus transformed cell lines were grown in tissue culture.
Experiments were conducted to compare rates of cholesterol
biosynthesis with levels of the regulatory enzyme HMG CoA
reductase and the membrane-located binding receptors for low
density lipoproteins (LDL) which mediate feedback control
in normal cells.

Elevations in cholesterol biosynthesis of 3-5 fold
from ^{14}C-acetate were observed both for Type II hypercholesterolemic cells and tumor cells. Correspondingly,
high levels of HMG CoA reductase were found in each cell
type. In contrast to hypercholesterolemic cells, however,
tumor cells exhibited efficient feedback control of this
enzyme when serum lipoproteins were added.

The absence of LDL receptors in hypercholesterolemic
cells (1), as evidenced by loss of ability to bind I^{125}-
labelled LDL, was confirmed. Monolayers of virus-transformed cells however, exhibited LDL binding characteristics
essentially the same as the normal fibroblasts.

These results confirm that the elevated cholesterol
biosynthesis reported for tumors *in vivo* (2) is also
exhibited by transformed cells *in vitro*. However, the
evidence of normal LDL-receptors and feedback control of
HMG CoA reductase in the tumor cells indicates that a lesion
in a control locus other than the membrane LDL-receptor
feedback loop is responsible for sterol overproduction in
human fibroblasts transformed by SV40 virus.

Supported by USPHS Grants #HL 05062, #CA 15356 and
NSF BMS 730085.

(1) Goldstein, J.L. and Brown, M.S., P.N.A.S. 71 788-792
    (1974).
(2) Siperstein, M., Gyde, A. and Morris, H., P.N.A.S. 68
    315-322 (1971).

VIRAL STIMULATION OF CHOLINE PHOSPHOTRANSFERASE IN SPLEEN MICROSOMES DURING PRODUCTION OF MALIGNANCY

W.E. Cornatzer, Dennis R. Hoffman and David Skurdal, Guy and Bertha Ireland Research Laboratory, Department of Biochemistry, University of North Dakota, Medical School, Grand Forks, N.D. 58202

The two enzymesystems involved in phosphatidyl choline (PC) biosynthesis in microsomes have been studied following viral infection during the production of cancer. PC is the major lipid in the membranes of plasma, endoplasmic reticulum and nucleus. Choline phosphotransferase (CPT) and phosphatidyl ethanolamine methyltransferase (PEMT) enzymatic activities (nmoles PC/min/mg protein) and total tissue enzymatic activity have been determined in spleen microsomes of Friend Virus and Rauscher Virus infected BALB/c male mice at 5, 10, 14 and 21 days following inoculation of the viruses. There is a significant marked stimulation of the CPT in viral infected spleen microsomes. There is very little stimulation of the PEMT activity during viral infection. The % increase over the control of the specific activity of CTP at 5, 10, 14 and 21 days of Rauscher Virus infection is 20%, 211%, 191% and 130% respectively. There is a 5-fold, 24-fold, 50-fold and 34-fold increase over controls for the total CPT activity in spleen microsomes at the same time intervals of viral infection. The % increase over the control of the specific activity of CPT at 5, 10, 14 and 21 days of Friend Virus infection is 201%, 149%, 200% and 19.6% respectively. A 39-fold, 51-fold, 61-fold and 24-fold increase over control for the total CPT activity at same time intervals of Friend Virus infection. Friend Virus tumor reached a maximum weight after 10 days of viral infection and represents a 21-fold increase over control. Rauscher Virus tumor growth is similar with a 19-fold increase over controls at 21 days. After 14 days of Friend Virus and/or Rauscher infection, the total microsomal phosphatidyl choline pool (μg lipid P/whole spleen) was increased 12 and 11-fold over controls respectively. The microsomal PC fractions (μ g lipid P/whole spleen) separated by $AgNO_3$ chromatography gave a % increase over controls; 884%, 885%, 996% and 1114% for PC fractions; 1, 2, 3 and 4 respectively after 14 days of Friend Virus infection and similar increases following Rauscher infection. The viral stimulation does not take place in liver tissue. The existence of two pathways of PC biosynthesis provided a source of different lecithin molecules for normal function and integrity of the membranes of the cell. Cancerogenic viruses stimulate this pathway and increase the concentration of PC. These phospholipid changes in the membranes of the cell that are due to viral infection may help to alter the characteristics of the cell membrane and thus be a part of the malignant process in the production of viral tumors.

L-Asparaginase with Antilymphoma Activity from Vibrio succinogenes

J. A. Distasio[1] and R. A. Niederman
Dept. of Microbiology, Rutgers University
New Brunswick, New Jersey

Homogeneous L-asparaginase with antilymphoma activity was purified from Vibrio succinogenes. An overall yield of 30-40% and a specific activity of 205 i.u. per mg of protein was obtained with this purification. The isoelectric point of the enzyme is 8.75. The K_m of the enzyme is 4.78×10^{-5} M. This is within the range of K_ms of known chemotherapeutically active enzymes. L-Asparaginase has a temperature optimum of approximately 45° and maintains at least 65% of its activity in the physiological temperature range.

The pH optimum is approximately 7.0 with 90% of the maximal enzyme activity maintained between pH 6.0 and 8.8. The enzyme catalyzes the hydrolysis of both the D- and L- isomers of asparagine with the activity in the presence of the L- isomer about 15 times that obtained with D- isomer. No other amide tested could be used as a substrate for this enzyme, except L-glutamine which was hydrolyzed less than 0.02% compared to L-aspargine. The high L-glutaminase activity of other chemotherapeutically active L-asparaginases has caused toxic and immunosuppressive side effects in patients being treated for acute lymphoblastic leukemia. L-Asparaginase from V. succinogenes was found to be a potent antilymphoma agent in C3H mice with transplanted 6C3HED lymphosarcomas. A group of 5 mice exhibited complete regression of tumors within 5 days after being injected with 2 i.u. of L-asparaginase per animal. The half life of the enzyme in the blood of C3H mice with transplanted 6C3HED lymphosarcomas was 26-31 hours.

[1] Present address: Laboratory of Virology, University of Miami School of Medicine, Box 520875, Miami, Florida

Research sponsored partially by the Charles and Johanna Busch Memorial Fund of Rutgers University.

MALATE-ASPARTATE SHUTTLE ACTIVITY IN SEVERAL ASCITES TUMOR LINES

W. V. V. Greenhouse and A. L. Lehninger,
Johns Hopkins University School of Medicine, Baltimore, MD

The activity of the malate-aspartate shuttle for reoxidation of cytoplasmic NADH by mitochondria was assessed in six lines of ascites tumors. The procedure entailed incubation of L-lactate with the cells and measurement of the pyruvate produced, which is a reflection of the cytosol NADH reoxidized. All the tumors examined showed reoxidation of cytoplasmic NADH (generated by the addition of lactate) which was nearly completely blocked by the transaminase inhibitor aminooxyacetate. That reoxidation of the NADH occurred via the respiratory chain and oxygen was shown by the addition of cyanide, rotenone, and antimycin A, which completely inhibited the formation of pyruvate from added L-lactate. Compounds which inhibit the facilitated entry of malate into mitochondria, such as butylmalonate, benzene tricarboxylate, and iodobenzylmalonate, also inhibited the accumulation of pyruvate from added L-lactate. The maximal rate of malate-aspartate shuttle activity was given when arsenite was added to inhibit mitochondrial oxidation of the pyruvate formed from lactate. The capacity for the reoxidation of cytoplasmic NADH via the malate-aspartate shuttle approaches 20% of the total respiratory rate of the various tumor cell lines tested. It is thus adequate to account for mitochondrial reoxidation of that fraction of glycolytic NADH not reoxidized by pyruvate.

Research supported by NIH Grant No. GM-05919 and NCI Contract No. NO1-CP-45610. Dr. W. V. V. Greenhouse is a Fellow of the Leukemia Society of America.

INHIBITION BY SERUM OF INTRACELLULAR DEGRADATION OF HUMAN CHORIONIC GONADOTROPIN (hCG)

R.O. Hussa and R.A. Pattillo,
Medical College of Wisconsin, Milwaukee, Wisconsin

This study was aimed at determining the mechanism by which 10% newborn calf serum enhances the secretion of hCG by human malignant trophoblast cells in vitro[1]. Although microgram amounts of hCG (in protein - free buffer) adsorbed to Falcon plastic culture flasks, addition of serum to the buffer both prevented and reversed the adsorption. Adsorption was not significant when the buffer consisted of medium in which cells had incubated for 24 hr prior to use. When flasks of trophoblast cells were incubated for 24 hr with ^{125}I-hCG + serum, no decrease was observed in the amount of ^{125}I-hCG precipitable with antibody against hCG. Trophoblast cultures were incubated for 1 day with ^{3}H-Val $\pm$ serum, and the radioactivity was determined by radioimmune precipitation (RIP) following addition of rabbit anti-hCG and goat anti-rabbit gamma globulin to a sample of the centrifuged medium. The ratio of ^{3}H (serum/no serum) in the immune precipitate from the medium was 1.4, whereas the respective ratio of total hCG (as measured by radioimmunoassay, RIA) was 27.5. Cells $\pm$ serum were incubated for 4 days with ^{3}H-Val + ^{14}C-glucosamine (^{14}C-Gm), then incubated in medium containing nonradioactive Val and Gm. The decrease in hCG s.a. (i.e., cpm ^{14}C or ^{3}H, RIP/IU hCG, RIA) was the same $\pm$ serum. When cells were incubated for 1 hr with ^{3}H-Val + ^{14}C-Gm, then incubated in the presence of nonradioactive Val and Gm, the maximum hCG s.a. occurred at the same time of chase for cells incubated either in the presence or absence of serum. These and other results suggested that serum increases hCG secretion by inhibiting intracellular degradation of hCG.

REFERENCES:

1) R.O. Hussa, M.T. Story and R.A. Pattillo, J. Clin. Endocrin. Metab. 40: 401, 1975.

Research sponsored by the National Institutes of Health, U.S.P.H.S., under Grant No. RO1-CA-14232-01A1.

THE INTERACTION OF ANTITUMOR DRUGS WITH FOLATE REQUIRING ENZYMES

D. W. Jayme, P. M. Harish Kumar, N. Appaji Rao, J. A. North, and J. H. Mangum, Graduate Section of Biochemistry, Department of Chemistry, Brigham Young University, Provo, Utah.

Little information is available on the interaction of antifolate compounds with enzymes of folate metabolism other than folate-H_2 reductase and thymidylate synthetase. However, since serine transhydroxymethylase, N^5,N^{10}-methylenefolate-H_4 reductase and methionine synthetase occupy pivotal positions in folate metabolism, their inhibition could profoundly affect normal and neoplastic tissues.

Serine transhydroxymethylase was purified 200-fold from pig kidney and 30-fold from L-1210 solid tumors grown subcutaneously in mice. The specific activity of this enzyme from pig kidney was about 10 times lower than that from the tumor tissue. The interaction of folate-H_4 with the kidney enzyme was allosteric ($S_{0.5}$ = 500µM; n = 3.9). However, the tumor enzyme was Michaelian (K_M = 300µM; n = 1). The enzyme found in liver homogenates obtained from normal and tumor-bearing mice also exhibited allosteric and Michaelian kinetics, respectively. Several antifolates inhibited both the enzymes with K_i values in the 1µM - 100µM range.

The N^5, N^{10}-methylenefolate-H_4 reductase was purified 1000-fold from pig kidney and 400-fold from L-1210 tumors. The specific activity of the homogenates from pig kidney and tumor were 0.10 and 0.04, respectively. Homofolate-H_4, N^5-methylhomofolate-H_4, methotrexate, aminopterin and dichloromethotrexate inhibited both the kidney and tumor enzymes. Dichloromethotrexate was a more effective inhibitor of the kidney enzyme (K_i = 4µM) than of the tumor enzyme (K_i = 80µM).

Methionine synthetase was also purified from pig kidney and tumor tissue. Cain's quinolinium and N^5-methylhomofolate-H_4 were potent inhibitors of this enzyme activity.

This work was supported by Contract No. 1-CM-43790 from Division of Cancer Treatment, NCI, NIH, DHEW.

STUDIES ON THE COLLAGENOLYTIC ACTIVITY OF METHYLCHOLAN-THRENE-INDUCED FIBROSARCOMAS IN MICE.

Kingsley R. Labrosse and Irvin E. Liener.
Department of Biochemistry
College of Biological Sciences
University of Minnesota
St. Paul, Minnesota 55108.

Methylcholanthrene-induced fibrosarcomas in C3H/Hej mice were found to possess two kinds of collagenolytic activity towards [^{3}H]-collagen, one having a pH optimum of 4.2 and the other pH 7.4. The activity observed at pH 4.2 was enhanced by cysteine and EDTA and inhibited by TLCK, TPCK, iodoacetate and p-chloromercuribenzoate, whereas the pH 7.4-active enzyme was inhibited by cysteine and EDTA and required calcium for activity. Both enzymes were inhibited by α_2-macroglobulin but neither one by α_1-antitrypsin. The pH 4.2 active enzyme was likewise capable of hydrolysing α-N-benzoyl-DL-arginine-β-napthylamide and thus appears to be identical in properties to cathepsin B1 which is of lysosomal origin. An electrophoretic examination of the products produced from collagen by the pH 4.2 and pH 7.4 active enzymes revealed that the former caused extensive degradation of collagen, whereas the latter yielded products of limited cleavage (β^A, α^A and α^B) characteristic of mammalian collagenases. When tumor cells are cultured in vitro, after the elimination of macrophages, leucocytes and lymphocytes from the preparation, the pH 7.4 activity appeared in the medium whereas the pH 4.2 activity remained bound to the tumor cells. An examination of the distribution of the collagenolytic activities in the solid tumor revealed both activities to be highest in the periphery or invasion zone of the tumor. A significant level of pH 4.2 activity could also be demonstrated in the necrotic center of the tumor but this region of the tumor was devoid of the pH 7.4 activity. The significance of these observations with respect to the invasive properties of tumors will be discussed.

This research was supported by USPH grants CA 16231 and CA 17774.

ACTIVITIES OF ENZYMES OF GLYCOLYSIS AND ADENINE NUCLEOTIDE METABOLISM DURING MURINE LEUKEMOGENESIS.

Lawrence A. Menahan and Robert G. Kemp

Department of Biochemistry, The Medical College of Wisconsin
Milwaukee, Wisconsin 53233

AKR mice develop lymphoid leukemia in high incidence at the age of 7 to 9 months. The causative agent is a vertically transmitted virus that apparently finds a suitable microenvironment in the thymus for the neoplastic transformation of lymphoid cells. Previous work from our laboratory has indicated significant changes occur in thymic cyclic AMP metabolism during the transition from the non-leukemic to leukemic state in the AKR mouse (1). These changes include a marked increase in adenylate cyclase in the preleukemic state that remained elevated during the leukemia. There was a slight increase in the tissue level of cyclic AMP in the preleukemic mouse followed by a sharp decrease in the thymic concentration of cyclic AMP in the lymphoma. In the present study, specific activities of four glycolytic enzymes (phosphofructo kinase, aldolase, pyruvate kinase and lactic dehydrogenase [LDH]) and adenosine deaminase, also a cytosolic enzyme, were elevated two to three fold in thymic lymphomas and a shift toward A_4 was observed in the LDH isozyme pattern. In the development of leukemia alterations in the specific activities of membrane bound enzymes were also observed. These included an increase in alkaline phosphatase and a decrease in 5'-nucleotidase, whose activity was specifically inhibited by α, β-methylene adenosine 5'-diphosphate. The most striking activity increase was that of membrane-bound cyclic AMP phosphodiesterase (PDE). Distribution studies indicated that the majority of cyclic AMP PDE activity measured at 1μM substrate sediments from a post-mitochondrial supernatant at 160,000 g where also a large share of the 5'-nucleotidase activity is found. Since 5'-nucleotidase is a marker enzyme for plasma membranes, it can be inferred that PDE activity measured at 1μM may be associated with plasma membrane. The increases in the membrane PDE observed in the thymic lymphoma may account for the sharply decreased cyclic AMP levels observed in this neoplasm.

REFERENCE
(1) Kemp, R.G., Hsu, P.-Y., and Duquesnoy, R.J., Cancer Res. 35:2440. 1975.

Research supported by Grant CA16539 awarded by the National Cancer Institute, DHEW.

THE ASSOCIATION OF A PROTEASE (PLASMINOGEN ACTIVATOR) WITH A SPECIFIC MEMBRANE FRACTION ISOLATED FROM TRANSFORMED CELLS

J.P. Quigley, Box 44, Department of Microbiology and Immunology, State University of New York, Downstate Medical Center, 450 Clarkson Avenue, Brooklyn, New York 11203

Proteases have been implicated as molecular regulators of growth control in normal and malignant cells. One protease, a secreted plasminogen activator (P.A.), has recently been shown to be elevated in tumor-virus transformed cultures and malignant cell lines. Its location and role intracellularly has not been examined in detail. This report quantitates the intracellular distribution of P.A. in transformed cells. Homogenates of Rous Sarcoma Virus transformed chick embryo fibroblasts (RSV-CEF) were fractionated by differential centrifugation followed by sucrose density gradient centrifugation. A series of marker enzymes, specific for different subcellular organelles, were used to analyze each fraction. The specific activity, recovery and % distribution of each marker enzyme was compared to that of P.A. activity. Analysis by differential centrifugation revealed that 80% of the P.A. was membrane-associated while only 10% was in the soluble fraction and 10% in the nuclear fraction. Further subfractionation yielded a membrane fraction that contained only 5-8% of the total cellular protein, was relatively free of mitochondria, microsomes and lysosomes, and contained the majority of the P.A. activity. The only other enzymes tested that exhibited a similar enrichment were the plasma membrane markers, 5' nucleotidase and Na^+K^+ ATPase. Further examination of this fraction indicated that P.A. was not released from its membrane association by hypotonic and hypertonic extraction and sonication. Granule bound enzymes, however, are released by these treatments. The use of different cell homogenization and sub-cellular fractionation procedures also indicated that P.A. is membrane-associated. The P.A. from a transformed cell line, hamster SV40, fractionated similarly to the RSV-CEF. The implied surface-membrane association of a transformation-dependent protease might have far-reaching implications for the role of proteases as biological regulators.

STRUCTURAL FEATURES OF S.TYPHIMURIUM LIPOPOLYSACCHARIDE (LPS) REQUIRED FOR ACTIVATION OF MONOCYTE TISSUE FACTOR.

F.R. Rickles and P.D. Rick, Depts. of Medicine and Microbiology, Univ. Conn. School of Med., VA Hosp., Newington, Conn.

Tissue factor (TF) activity has been demonstrated in suspensions of several types of mammalian cells including fibroblasts, granulocytes, monocytes and endothelial cells. Recent studies indicate that TF exists in an inactive form in association with the cell surface. These findings suggest that a specific activation process is required to generate procoagulant activity. Since TF is an ubiquitous coagulation protein it is important to determine the nature of this activation process. The cell envelope LPS of gram-negative bacteria are potent activators of granulocyte and monocyte TF. The structural features of LPS required for monocyte TF activation have been studied employing LPS derivatives and incomplete (rough) LPS from mutant strains of S.typhimurium. Mononuclear cells were obtained from heparinized human blood by isopycnic centrifugation. Monocytes comprised 10-30% of the cells and were responsible for the generation of TF in culture; lymphocytes and granulocytes had no apparent role as we have previously shown (Blood, Vol. 46, in press). Results of these experiments are summarized in the table below. LPS from a heptoseless strain of S.typhimurium (G30-A) lacks carbohydrate residues distal to the ketodeoxyoctonate (KDO)-lipid A region of the molecule. This material had the same capacity for activating TF as the complete (smooth) LPS. Removal of the KDO residues from G30-A LPS by mild acid hydrolysis failed to diminish the ability of this material to activate TF. Mild alkaline hydrolysis of the lipid A, however, destroyed this property. It is suggested that the integrity of lipid A is necessary for the activation of monocyte TF by LPS.

MATERIAL (1 ug/ml)	INTACT LPS	G30-A	LIPID A	ALK.HYD. LIPID A
UNITS of $TF/1.5 \times 10^6$ cells*	49.9 ± 4.3	61.9 ± 10.4	44.0 ± 3.5	< 1.0

*$\pm$1 SEM.

ROLE OF PLASMINOGEN ACTIVATOR IN GENERATION OF MIF-LIKE ACTIVITY BY SV3T3 CELLS

R.O. Roblin, M.E. Hammond, P.H. Black and H.F. Dvorak -- Depts. of Microbiology and Molecular Genetics, Medicine, and Pathology, Harvard Medical School and Massachusetts General Hospital, Boston, MA 02114

Culture medium from SV-40 transformed 3T3 cells (but not 3T3 cells) cause guinea pig peritoneal exudate (PE) cells to decrease their rate of migration from capillary tubes and to lose a densely staining cell surface material (1). Media from other cell types transformed by either DNA or RNA tumor viruses similarly contains a migration inhibitory factor (MIF)-like activity (2). Isolation and partial purification of a factor from SV3T3 cells which can generate MIF-like activity has shown: 1) that overnight incubation of serum free harvest fluids (HF) from SV3T3 cultures with medium containing 15% fresh guinea pig serum (GPS) generates an MIF-like activity; 2) that the generation of both the MIF-like activity by SV3T3 HF and the HF's plasminogen activator activity, can be inhibited by prior incubation of the HF with ^{3}H-diisopropylfluorophosphate (^{3}H-DFP); 3) that 3H-DFP reaction with SV3T3 HF labels predominantly a component of MW approx. 50,000 by SDS-polyacrylamide gel electrophoresis; 4) that the ^{3}H-DFP labeled component, the plasminogen-dependent fibrinolytic activity, and the ability to generate MIF-like activity upon overnight incubation with medium containing 15% GPS, co-chromatograph on Sephadex G200 columns run in Glycine-HCl buffer, pH 3.3. These results and other data indicate that the plasminogen activator elaborated by SV3T3 cells is reponsible for generating the MIF-like activity observed in cultures of SV3T3 cells by interacting with a component or components of GPS. Since increased secretion of cellular plasminogen activator(s) frequently accompanies viral transformation, our results suggest that plasminogen activator(s) may generate the MIF-like activity associated with virus-transformed cells (1,2) and possible also the MIF-like activity observed in the serum of patients with lymphoproliferative diseases (3).

(1) Hammond, M.E., Roblin, R.O., Dvorak, A.M., Selvaggio,S.S., Black, P.H. and Dvorak, H.F. Science 185, 955 (1975).

(2) Poste, G. Cancer Research 35, 2558 (1975).

(3) Cohen, S., Fisher, B., Yoshida, T. and Bettigole, R.E. New England Journal of Med. 290, 882 (1975).

PROTEOLYSIS AND CYCLIC AMP LEVELS IN CELL CULTURE

W.L. Ryan, M.L. Heidrick, and G.L. Curtis,
University of Nebraska College of Medicine, Omaha, Nebraska

Limited proteolysis with 1:300 trypsin or fibrinolysin initiates mitosis and division in confluent cell cultures (1). Associated with the increase in cell division is a considerable decrease in cellular cyclic AMP. Similar depressions in cyclic AMP following treatment with 1:300 trypsin have been described by others (2). Examination of cyclic AMP levels of cells throughout the growth cycle of normal rat embryo cells (F111) revealed large increases in cyclic AMP shortly after plating the cells (3). This increase in cyclic AMP occurs only when the cells used for plating are removed with trypsin. Removal by scraping does not increase the cyclic AMP levels. These data suggest trypsin treatment may increase as well as decrease cyclic AMP in cells.

In order to determine the mechanism by which trypsin alters cyclic AMP levels, adenylate cyclase of F111 cells was assayed following treatment with 1:300 or crystalline trypsin. Both enzyme preparations stimulate adenylate cyclase and the degree of stimulation is affected by trypsin concentration and length of treatment. The stimulation by trypsin can be blocked with soybean inhibitor. Fluoride-stimulated adenylate cyclase was further stimulated by trypsin treatment which suggests that the catalytic portion of the enzyme is being altered. The PGE_1 response is still present in the trypsin-treated enzyme indicating that the receptor is not altered by the treatment.

Investigation of the response of intact F111 cells to proteolysis indicates that the cyclic AMP response to 1:300 trypsin is pH dependent and this provides a possible explanation for the contradictory data described.

REFERENCES

(1) Burger, M.M., Bombik, B.M., Breckenridge, B.M. and Sheppard, J.R., Nature New Biol. 239, 161, 1972.

(2) Otten, J. Johnson, G.S. and Pastan, I., J. Biol. Chem. 247, 7082, 1972.

(3) Ryan, W.L. and Curtis, G.L., Chemical Carcinogenesis and Cyclic AMP in Role of Cyclic Nucleotides in Carcinogenesis, 6, 1, 1973.

SPECIFICITY OF DNA-DEPENDENT RNA POLYMERASE ACTIVITIES IN RABBIT BONE MARROW ERYTHROID CELL NUCLEI

M.K. Song and J.A. Hunt,
Department of Genetics, University of Hawaii School of Medicine, Honolulu, Hawaii

Two main types of DNA-dependent RNA polymerases have been isolated from rabbit bone marrow erythroid cell nuclei according to the method of Roeder and Rutter[1]. One is primarily nucleolar (RNA polymerase I) and the other primarily nucleoplasmic (RNA polymerase II). These two enzymes differed in their relative activation by manganese and magnesium, inhibition by a-amanitin, and ammonium sulfate optima. The relative activity of RNA polymerase I compared to RNA polymerase II was increased when calf thymus DNA was used as template instead of rabbit DNA. Both RNA polymerase activities increased three fold when calf thymus DNA was heat denatured. RNA polymerase activities were also characterized by the size and base composition of RNA transcribed _in vitro_ from rabbit bone marrow chromatin. In the presence of magnesium, RNA polymerase I synthesized RNA with a G-C content of about 60%. RNA polymerase II, in the presence of manganese, synthesized RNA with a lower G-C content (45-54%). Sedimentation analysis of the products showed that RNA polymerase II synthesized RNA with a size of about 45S in the presence of manganese and 35S in the presence of magnesium. RNA polymerase I synthesized RNA smaller than 28S. RNA polymerase I products synthesized in the presence of magnesium were slightly larger than those synthesized in the presence of manganese. Our results suggest that the two RNA polymerases isolated from rabbit bone marrow erythroid cell nuclei have distinct properties with respect to the nucleotide sequences which they transcribe. This property may in part lead to the specific manner in which genes are transcribed _in vivo_.

REFERENCES

(1) R.G. Roeder and W.J. Rutter, Biochemistry 9:2543-2553 (1970).

Research supported by Grant No. GM19076 from the U.S. Public Health Service.